Low-Carbon Development for Mexico

Low-Carbon Development for Mexico

Todd M. Johnson
Claudio Alatorre
Zayra Romo
Feng Liu

THE WORLD BANK
Washington, DC

ISBN: 978-0-8213-8122-9
eISBN: 978-0-8213-8123-6
DOI: 10.1596/978-0-8213-8122-9

Library of Congress Cataloging-in-Publication Data

Low-carbon development for Mexico / Todd M. Johnson... [et al.].
  p. cm.
  Includes bibliographical references and index.
  ISBN 978-0-8213-8122-9 — ISBN 978-0-8213-8123-6 (electronic)
1. Energy policy—Mexico. 2. Power resources—Mexico. 3. Carbon dioxide
mitigation—Mexico. I. Johnson, Todd (Todd Milo), 1956–
  HD9502.M62L69 2009
  363.738'7460972—dc22
                                                                    2009035879

Cover photograph: Andres Balcazar, ©iStockphoto.com/abalcazar
Cover design: Critical Stages

Contents

Preface

One of the most compelling reasons for pursuing low-carbon development is that the potential impacts of climate change are predicted to be severe, for both industrial and developing countries, and that reducing greenhouse gas emissions can reduce the risk of the most catastrophic impacts. The challenge of reducing emissions is sobering: leading scientific models indicate that limiting the rise in global mean temperatures to less than 2°C will require that global greenhouse gas emissions peak within the next 10–15 years and then fall by 2050 to levels about 50 percent lower than in 1990. Although many countries recognize the need to curtail carbon emissions, there is considerable uncertainty about how much this will cost in individual countries, what measures can be undertaken in both the short and longer term, and how cost-effective specific interventions are in reducing emissions.

"Low carbon" is quickly entering the lexicon of development, adding an important climatic dimension to the concept of economic sustainability. *Low-Carbon Development for Mexico* provides an economywide analysis of low-carbon options for mitigating greenhouse gas emissions in Latin America's largest fossil fuel–consuming country. The study is the first of several low-carbon studies to be produced by the World Bank in key developing and middle-income countries.

Mexico was a logical choice for a low-carbon study for several reasons. At the international level, it has demonstrated strong commitment to global actions to reduce greenhouse gas emissions, as reflected in its proactive stance in global climate discussions and the aggressive emission reduction target it announced at the United Nations Climate Change Conference in Poznan in 2008. At home, Mexico recently published the *Programa Especial de Cambio Climático* (PECC), which sets out a broad program to address the impacts of climate change in Mexico and to reduce greenhouse gas emissions across all sectors.

This volume, intended to complement the PECC and other Mexican studies, presents the results of a two-year effort by a team of Mexican and international researchers to identify and evaluate high-priority measures for reducing greenhouse gas emissions. The study makes use of two important tools for undertaking low-carbon assessments. The first is an economic methodology for estimating the costs of interventions across sectors. This methodology allows, for example, the costs of reducing emissions from introducing more efficient residential refrigerators to be compared with those achieved through afforestation or reforestation programs. A second tool is an integrated economic and emissions model that keeps track of annual emissions as well as needed investment costs over the coming two decades.

The need to reduce emissions associated with energy production and consumption—including from transport and power generation—is often at the heart of discussions about low-carbon development. The fastest emissions growth in Mexico over the past three decades has occurred because of rising energy consumption in the road transportation sector, and the growth in private automobiles and light trucks is expected to continue to fuel this growth in the future. This study presents new research on low-carbon interventions in the transport sector, including measures to improve the efficiency of both new and used vehicles as well as measures to improve urban transportation. Because a large percentage of transportation energy use occurs in Mexico's cities, there is significant potential for lowering greenhouse gas emissions by modifying the spatial organization of cities and improving the availability of public transportation infrastructure. Although major changes in urban design will take time to develop, other measures—such as investing in BRT-type systems, strengthening public transportation, and reorganizing freight transport systems—can be implemented in the near term.

This study analyzes a range of energy efficiency options available in Mexico, including supply-side efficiency improvements in the electric power and oil and gas industries, and demand-side electricity efficiency measures addressing high-growth energy-consuming activities, such as air conditioning and refrigeration. It also evaluates a range of renewable energy options that make use of the country's vast wind, solar, biomass, hydro, and geothermal resources.

But low-carbon development is not only about energy production and consumption. In Mexico one of the most important sources of greenhouse gas emissions continues to be deforestation. The rate of deforestation has fallen steadily in Mexico over the past decades. Expanded programs for forest management, wildlife management, and efforts to increase the stock of forests can provide needed employment in rural areas and help make Mexican forests net absorbers of CO_2 in the coming years.

A fundamental question often asked about low-cost mitigation options is why they are not already being undertaken. As the study shows, the availability of commercial technology and even low financial costs is often not enough to overcome barriers related to institutional and knowledge gaps, regulatory and legal constraints, or societal norms. Inability to surmount these "transactions costs" is typically at the root of the problem of why supposedly low-cost actions are not undertaken. To partially overcome this dilemma, one of the explicit criteria used in this study for identifying low-carbon measures was that they had already been implemented on some scale in Mexico or in a similar economy outside of Mexico. In order to mainstream low-carbon development, a package of new stimuli will be needed, including public and consumer education and training, public demonstrations, standards and regulations, and financial incentives.

The next few years will be critical for enacting a serious international climate mitigation program, beginning with major industrial countries and quickly involving large developing countries. A number of mitigation studies have looked at the longer term, many of them focusing on the promise of new technologies to achieve significant reductions in carbon emissions. Although new technologies will be critical to meeting the longer-term emissions reduction goals needed to avoid the most severe impacts of climate change, many promising low-carbon technologies will not be commercially available for more than a decade, during which time the world will lose valuable degrees of freedom in stabilizing atmospheric concentrations, if short-term options have not been simultaneously and vigorously pursued. One of the explicit objectives of this study was to identify a range of options that could contribute to meaningful emissions reductions over the next two decades and that could begin almost immediately. As new technologies are developed and the costs of current technologies fall, the range of options for low-carbon development will become even broader.

Although this study focuses on Mexico, many of the low-carbon options presented—such as specific energy-efficiency and renewable energy technologies and urban transport or forestry programs—are likely to be applicable to other countries. It is our hope that both the methodologies and the findings presented in this volume will be of use to Mexico and other countries as they seek to define and implement low-carbon development.

Laura Tuck, Director
Sustainable Development Department
Latin America and the Caribbean Region
The World Bank

About the Authors

Todd M. Johnson is a lead energy specialist in the Sustainable Development Department of the Latin America and the Caribbean Region of the World Bank. Since joining the Bank in 1991, he has worked on a variety of energy- and environment-related topics, including acid rain control and climate change. He has coauthored numerous articles and reports, including *China: Issues and Options in Greenhouse Gas Emissions Control* (1994), *Climate Change Mitigation in the Urban Transport Sector* (2003), and *Residential Electricity Subsidies in Mexico: Exploring Options for Reform and for Enhancing the Impact on the Poor* (2009). He holds a PhD in economics from the University of Hawaii.

Claudio Alatorre is an independent consultant with expertise in energy transition (energy efficiency, renewable energy, sustainable transport) and the design and implementation of enabling policy and institutional frameworks. He has worked with academic institutions, nongovernmental organizations, private firms, the media, multilateral and bilateral agencies, and government institutions in Mexico and other countries. He holds a PhD in engineering from Warwick University, the United Kingdom.

Zayra Romo is a power specialist in the Sustainable Development Department of the Latin America and the Caribbean Region of the World Bank. She provides technical and financial analysis for generation and transmission infrastructure projects. Before joining the Bank, she was a technical analyst for Électricité de France, where she worked on improving the performance of power plants in Mexico. She holds an MSc in energy conversion from the University of Offenburg, Germany.

Feng Liu is a senior energy specialist in the Energy Sector Management Assistance Program (ESMAP), a multidonor partnership administered by the World Bank. For the past 10 years, he has been involved in the development and implementation of energy-efficiency and renewable energy investment projects in the East Asia and Pacific Region, particularly in China, Indonesia, and Mongolia. Before joining the Bank, he spent five years conducting energy analysis and policy research at Lawrence Berkeley National Laboratory in California. He holds a PhD in environmental economics from Johns Hopkins University.

Acknowledgments

This study was initiated by the World Bank as one of six low-carbon studies to be carried out in developing and middle-income countries. The study concept was discussed with Mexican government authorities in 2007; the decision to undertake the study was endorsed by the ministries of energy (SENER), environment (SEMARNAT), and finance (SHCP).

This study was supported by the World Bank through funds made available from the Sustainable Development Network for regional climate change activities and through support from the United Nations Development Programme (UNDP)/World Bank Energy Sector Management Assistance Program (ESMAP). The financial assistance of the government of the United Kingdom (through the Department for International Development [DFID]) through ESMAP is gratefully acknowledged.

This report was prepared by Todd M. Johnson (task manager and lead author), Zayra Romo, and Feng Liu, all with the World Bank, and by Claudio Alatorre, a consultant. Other contributors included the following:

- **Agriculture, forestry, and bioenergy:** Javier Aguillón, Marcela Olguín-Álvarez, Tere Arias, Víctor Berrueta, Guillermo Colunga, Jorge Etchevers, Carlos Alberto García, Adrián Ghilardi, Rocío Gosch, Gabriela Guerrero, Ben de Jong, Omar Masera, Mauricio Pareja, Manuela Prehn, Oliver Probst, Enrique Riegelhaupt, Emilio de los Ríos, and Juan Angel Tinoco. Representing different organizations in Mexico, these experts are members of the Red Mexicana de Bioenergia (REMBIO) (Mexican Network for Bioenergy).
- **CGE modeling:** Roy Boyd, Ohio University, and María Eugenia Ibarrarán, Universidad Iberoamericana Puebla.
- **Electric power:** Myriam Cisneros, Jorge Gasca, Moisés Magdaleno, Elizabeth Mar, Luis Melgarejo, and Esther Palmerín, Instituto Mexicano del Petróleo.

- **Coordination, modeling, energy efficiency, and oil and gas:** Odón de Buen, Emmanuel Gómez-Morales, Genice Kirat Grande, Jorge M. Islas-Samperio, Paloma Macías-Guzmán, Fabio Manzini, María de Jesús Pérez-Orozco, and Mario Alberto Ríos-Fraustro, Centro de Investigación en Energía, Universidad Nacional Autónoma de México; and independent consultants.
- **Cost-benefit analysis:** Carlos E. Carpio, Tomás Hasing, James B. London, Matías Nardi, William A. Ward, Gary Wells, and Samuel Zapata, Clemson University.
- **Transport:** Amílcar López, Jorge Macías-Mora, Hilda Martínez, Gabriela Niño, Luis Sánchez-Cataño, and Juan Sebastián Pereyra, Centro de Transporte Sustentable de México, A.C.

Members of the study team from the World Bank included Benoit Bosquet (land use and forestry), Francisco Sucre (oil and gas), and Jas Singh (energy efficiency).

The report benefited from comments and suggestions from Ricardo Ochoa, Secretaría de Hacienda y Crédito Público; Fernando Tudela and Juan Mata, Secretaría de Medio Ambiente y Recursos Naturales (SEMARNAT); Adrián Fernández, Instituto Nacional de Ecología (INE); Veronica Irastorza, Francisco Acosta, and Diego Arjona, Secretaría de Energía (SENER); Emiliano Pedraza, Comisión Nacional para el Uso Eficiente de la Energía (CONUEE); Vicente Aguinaco, Comisión Federal de Electricidad (CFE); Carlos De Regules, Petróleos Mexicanos (Pemex); and anonymous reviewers from a number of Mexican institutions.

The report was prepared under the direction of Laura Tuck, Axel van Trotsenburg, and Philippe Charles Benoit. World Bank staff who provided comments and suggestions as part of the review process included Roberto Aiello, Jocelyne Albert, Amarquaye Armar, Juan Carlos Belausteguigoitia, Pablo Fajnzylber, Marianne Fay, Charles M. Feinstein, Christophe de Gouvello, Ricardo Hernandez, Richard Hosier, Irina Klytchnikova, Kseniya Lvovsky, John Nash, Paul Procee, John Allen Rogers, Gustavo Saltiel, Ashok Sarkar, Gary Stuggins, Natsuko Toba, and Walter Vergara.

Production assistance from Janina Franco, Aziz Gokdemir, Barbara Karni, Nita Congress, and Michael Alwan is gratefully acknowledged.

Abbreviations

APEC	Asia-Pacific Economic Cooperation
bcm	billion cubic meters
BOE	barrel of oil equivalent
BRT	bus rapid transit
CAFE	corporate average fuel economy
CDM	Clean Development Mechanism
CFE	Comisión Federal de Electricidad (Federal Electricity Commission)
CGE	computable general equilibrium
CICC	Comisión Intersecretarial de Cambio Climático (Intersecretarial Commission on Climate Change)
CO_2	carbon dioxide
CO_2e	carbon dioxide equivalent
CONAE	Comisión Nacional para el Ahorro de Energía (National Commission for Energy Savings)
CONAGUA	Comisión Nacional del Agua (National Water Commission)
CONUEE	Comisión Nacional para el Uso Eficiente de la Energía (National Commission for the Efficient Use of Energy)
DDG	dried distillers grain
E&D	exploration and development
ENACC	Estrategia Nacional de Cambio Climático (National Climate Change Strategy)
ESCO	energy service company
FCC	fluidized catalytic cracking
FIDE	Fideicomiso para el Ahorro de Energía Eléctrica (Fund for Electricity Savings)
GDP	gross domestic product

GHG	greenhouse gas
GJ	gigajoule
GJ/t	gigajoules per ton
GW	gigawatt
GWh	gigawatt-hour
I&M	inspection and maintenance
IEA	International Energy Agency
INEGI	Instituto Nacional de Estadística, Geografía e Informática
IPCC	Intergovernmental Panel on Climate Change
IPP	independent power producer
kWh	kilowatt-hour
LEAP	Long-range Energy Alternatives Planning
LNG	liquefied natural gas
LPG	liquefied petroleum gas
LULUCF	land use, land-use change, and forestry
LyFC	Luz y Fuerza del Centro
mbd	million barrels per day
mcfd	million cubic feet per day
MEDEC	México: Estudio sobre la Disminución de Emisiones de Carbono (Mexico Low-Carbon Development Study)
MEPS	minimum energy performance standards
MJ	megajoule
Mm^3	million cubic meters
Mt	million tons
Mtoe	million tons of oil equivalent
MW	megawatt
NO$_X$	nitrogen oxides
PECC	Programa Especial de Cambio Climático (Special Climate Change Program)
Pemex	Petróleos Mexicanos
PIDIREGAS	Proyectos de Impacto Diferido en el Registro de Gasto (Projects with Differed Expenditure Impact)
PJ	petajoule
PM2.5	fine particles 2.5 micrometers in diameter or smaller
PM10	fine particles 10 micrometers in diameter or smaller
PND	Plan Nacional de Desarrollo (National Development Plan)
REDD	reducing emissions from deforestation and forest degradation
SEMARNAT	Secretaria de Medio Ambiente y Recursos Naturales
SENER	Secretaría de Energía (Ministry of Energy)
SO$_2$	sulfur dioxide
SO$_4$	sulfate
SUV	sport utility vehicle
t	ton

TDM	tons dry matter
TWh	terawatt-hour
UMA	Unidades de Manejo para la Conservación de la Vida Silvestre (management units for wildlife conservation)
UNFCCC	United Nations Framework Convention on Climate Change
WTP	willingness to pay

All references to $ are US$.
All tons are metric tons.

Overview

Mexico's Special Climate Change Program—the Programa Especial de Cambio Climático (PECC), published in August 2009—sets Mexico's long-term climate change agenda, together with medium-term goals for adaptation and mitigation. This study—known as México: Estudio sobre la Disminución de Emisiones de Carbono (MEDEC)—is intended to contribute to the implementation of that long-term climate change agenda.

The study evaluates the potential for reducing greenhouse gas emissions in Mexico over the next 20 years. It evaluates low-carbon interventions across key emission sectors in Mexico using a common methodology. Based on the interventions evaluated, it develops a low-carbon scenario through 2030.

Benefits of Moving to a Low-Carbon Economy

Reducing greenhouse gas emissions is critical in Mexico, not only to address climate change but also to facilitate economic development, a key emphasis of the country's climate change agenda. Moving to a low-carbon economy could benefit Mexico in at least four ways:

- Because it is likely to suffer disproportionately from the impacts of climate change (drought, sea level rise, increased severity of tropical storms), Mexico has a strong interest in becoming a leading participant in an international agreement to cap emissions.
- Numerous "no-regrets" low-carbon interventions (interventions that have positive economic rates of return and should be undertaken irrespective of climate change considerations) can contribute substantially to economic development in Mexico.
- Many low-carbon interventions have important co-benefits for Mexico, including the enhanced energy security associated with energy

efficiency (on both the supply and demand sides) and renewable energy projects; the human health benefits from transport and other inventions that reduce local air pollutants; and the environmental protection benefits that can be achieved through forestry and natural resource management, waste-reduction programs, and reduced emissions of local pollutants from energy facilities.

- Countries that pursue low-carbon development, including the transfer of financial resources through the carbon market and new public programs that support climate change mitigation, are likely to reap strategic and competitive advantages.

Mitigation Options, by Sector

The MEDEC study evaluated low-carbon interventions in five sectors: electric power, oil and gas, stationary energy end-use, transport, and agriculture and forestry. Three criteria were used to select interventions:

- Interventions had to have substantial potential for reducing greenhouse gas emissions. The threshold for including an intervention was 5 million tons of CO_2–equivalent (Mt CO_2e) over the 2009–30 implementation period.
- Interventions had to have low economic and financial costs. First priority was given to no-regrets interventions. A second tier of projects—with carbon costs of $25/t or less—was also included.
- Interventions had to be feasible in the short or medium term. Ensuring that this criterion was met required investigation of information, regulatory, and institutional barriers that are keeping low-carbon interventions from being adopted on a large scale. Feasibility was first determined by sectoral experts; it was then discussed with government officials and international experts. All MEDEC interventions have already been implemented, at least on a pilot level, in Mexico or in countries facing similar conditions. Some interventions face barriers in the short term (next five years), but the barriers preventing their adoption are believed to be surmountable in the medium term.

Electric Power

The demand for electric power in Mexico has been growing faster than gross domestic product (GDP) over the past several decades, and this trend is likely to continue. Under a baseline scenario, meeting the increasing demand for power would increase total CO_2e emissions from power generation by 230 percent between 2008 and 2030 (from 142 Mt CO_2e to 322 Mt CO_2e). Both coal- and gas-fired power generation would increase under this scenario, with coal accounting for 37 percent of new installed capacity and natural gas accounting for 25 percent.

Assuming a net cost of CO_2e of as little as $10/ton, additional low-carbon energy technologies—small hydro, wind, biomass, geothermal, cogeneration (that is, the combined generation of heat and electricity in the same facility)—could replace much of the fossil fuel generation (principally coal but also natural gas) in the baseline scenario. Under the low-carbon MEDEC scenario, the share of power generated by coal would decline from 31 percent to 6 percent, and the contribution of low-carbon technologies would increase substantially, rising from 1.4 percent to 6.0 percent for wind, 2 percent to 11 percent for geothermal, 0.1 percent to 8.0 percent for biomass, and 14 percent to 16 percent for hydro. At net costs that are less than current marginal costs of power generation in Mexico, cogeneration would provide 13 percent of new power capacity under the low-carbon scenario. Abatement costs were calculated by comparing the net costs (including capital, energy, and operations and maintenance costs) of each low-carbon technology with the costs of the displaced coal and natural gas capacity.

Several policy and regulatory changes are needed to expand the share of renewable energy and energy efficiency in the power sector. Although the costs of wind generation in Mexico are among the lowest in the world—because of the high-quality wind resources in the isthmus of Tehuantepec, where some new wind projects are being developed—the country's enormous wind resources have not been widely developed. Factors inhibiting the development of wind and other renewables include low planning prices and the absence of externalities that Mexico's federal electricity commission, Comisión Federal de Electricidad (CFE), has historically assumed for new fossil fuel–based power generation; the lack of recognition of the portfolio effect in power planning, which would increase the share of renewable energy interventions based on their lower fuel risk; and the inability to adjust procurement procedures to the particularities of renewable energy projects. New contracting procedures are needed for cogeneration and other small-scale projects to reduce the risks and transaction costs of small power producers.

Oil and Gas

There is significant potential to reduce greenhouse gas emissions in Mexico's oil and gas sector through both no-regrets and low-cost interventions. In particular, significant cogeneration potential at Pemex facilities could provide more than 6 percent of Mexico's current installed power capacity.

Specific interventions that can reduce greenhouse gas emissions and have good economic rates of return include reducing gas distribution leakage; increasing efficiency at Pemex oil, gas, and refining facilities; and realizing the cogeneration potential at Pemex's six refineries and four petrochemical plants. Developing this potential will require a regulatory framework that enables and encourages the sale of excess energy and capacity to the electricity grid.

Despite their excellent rates of return, investments in cogeneration and reductions in gas leakage are less attractive to Pemex than investments in oil exploration and development. Financing of investment is also difficult, for two reasons. First, Pemex's high debt—the highest of any oil company in the world in 2007—has made it difficult to tap commercial credit markets at reasonable terms. This problem will become even more difficult given the recent international financial crisis, despite the recent passage of oil industry reform measures. Second, although the oil industry accounts for only about 6 percent of GDP, oil revenues account for more than one-third of Mexico's federal budget. This constrains the government from taking measures that reduce tax payments from Pemex in the short term. Measures to allow contracting with the private sector to tap cogeneration and reduce gas flaring and leakage could reduce the need for public investment.

Although the MEDEC scenario reduces the demand for natural gas compared with the baseline, MEDEC and other recent studies foresee a major increase in the absolute amount of natural gas consumption. The success of the government's plan to expand natural gas production is therefore extremely important.

Energy End-Use

Electricity demand in Mexico has grown by more than 4 percent a year since 1995. Managing this growth through energy-efficiency measures in the end-use sectors will be critical to mitigating greenhouse gas emissions.

More than half of industrial energy use occurs in three subsectors: cement, iron and steel, and chemicals and petrochemicals. Many of Mexico's large-scale basic materials industries, including iron, steel, and cement, are among the most efficient in the world. The problem is that a large portion of the industrial sector is made up of small and medium enterprises that often use old equipment and lack access to technical know-how and financing for upgrades. These companies have relatively high energy intensity. The main sources of energy savings in the industrial sector come from energy-efficiency improvements in motor and steam systems and in kilns and furnaces, as well as from cogeneration—for which more than 85 percent of the industrial potential has not been utilized.

Air conditioning, refrigeration, and electronics are expected to be the main growth areas of residential electricity demand in Mexico. Air conditioner saturation rates in Mexico were about 20 percent in 2005—far lower than the 95 percent rates in regions of the United States with similar cooling-degree days. The saturation rate of refrigerators is relatively high in Mexico, at 82 percent in 2006, but it is still expected to grow considerably. Recent efforts to promote compact fluorescent lamps notwithstanding, incandescent lamps account for about 85 percent of in-use residential light bulbs in Mexico, indicating large potential for scaling up replacement efforts. There is also significant mitigation potential through solar water heating in urban areas and improved fuelwood cookstoves in rural areas.

Policies to improve efficiency in the residential, commercial, and public sectors—including tightening and enforcing efficiency standards for lighting, air conditioning, refrigeration, and buildings—will be critical to limit greenhouse gas emissions. As the analysis shows, the investment required in all electricity-efficiency interventions is significantly less than the investment in power plants that would otherwise be needed.

Transport

Transport is the largest and fastest-growing sector in terms of both energy consumption and greenhouse gas emissions in Mexico, with road transport accounting for about 90 percent of the sector's CO_2e emissions. Between 1996 and 2006, Mexico's vehicle fleet nearly tripled, increasing from 8 million to more than 21 million vehicles. Energy use by road transport increased more than fourfold between 1973 and 2006. The importation of used vehicles from the United States has been an important factor behind the growth of the vehicle fleet, which has also led to an increase in the average fleet age and concerns about low gas mileage and high emissions of air pollutants.

A number of interrelated interventions that reduce greenhouse gas emissions in the transport sector were evaluated. They included increasing the density of urban development, raising energy-efficiency standards for new vehicles, optimizing transportation routes, creating a bus rapid transit (BRT) system, encouraging nonmotorized transport, mandating the inspection and maintenance of in-use vehicles in major cities, imposing import restrictions on vehicles through inspection, coordinating road freight, and promoting freight trains.

Given the historical and projected urbanization pattern in Mexico, urban transport and related land-use planning issues will be a critical component of overall energy usage by the transport sector and associated emissions. The analysis reveals the importance of addressing transport issues in an integrated and programmatic approach rather than as individual measures. The interventions with the largest potential that are most cost-effective are those that increase the percentage of trips by public transportation and improve the efficiency of the vehicle fleet. Increasing the use of public transportation—including through private concessions—will require the development of mechanisms that integrate public transportation and urban development efforts by both federal and municipal governments. Promoting more sustainable transport policies can provide numerous co-benefits in addition to climate change mitigation, including reductions in traffic congestion (and the associated time savings per trip) and improvements in public health as a result of reduced air pollution.

Agriculture and Forestry

Agriculture and forestry is one of the key sectors in which greenhouse gas emissions can be reduced in Mexico. The MEDEC interventions are based on a geographical model that determined the areas that can be devoted to various rural activities while minimizing possible negative impacts on food production and biodiversity conservation. The interventions in forestry—including reforestation, commercial plantations, and measures to reduce emissions from deforestation and forest degradation (REDD)—account for 85 percent of the proposed mitigation in the agriculture and forestry sector. They are among the most important mitigation options for Mexico. The interventions in this sector that have the highest benefits are those that both substitute fossil fuel use through the sustainable production of biomass energy and reduce deforestation and forest degradation.

Many of the forestry interventions have unquantified environmental benefits, such as soil conservation, improvements in water quality, and preservation of ecosystems, in addition to the quantified benefits of income generation and employment for rural communities. Successful expansion of forestry sector interventions in Mexico depends on institutional changes in forest management, improved public financing mechanisms, and the development of a market for sustainable forest products.

Cost-effective measures for reducing greenhouse gas emissions from the agricultural sector are more limited, partly because of the lack of research and development on low-carbon measures. However, minimum tillage for maize production—which requires less energy and appears to facilitate soil carbon sequestration—appears to be a promising technology.

Sugarcane ethanol has significant greenhouse gas reduction potential, although the productivity of sugarcane production in Mexico is currently low (production costs are significantly above world market prices of sugar). Other liquid biofuels interventions—ethanol from sorghum and biodiesel from palm and jatropha—are estimated to have limited reduction potential without impinging on land use for food crops, forests, or conservation lands. All liquid biofuels options have positive net economic costs when compared with the opportunity cost of selling the feedstocks for food or other nonfuel uses.

Emissions Reductions Associated with a Low-Carbon Scenario

The baseline scenario was generated using the LEAP (Long-range Energy Alternatives Planning) model, based on macroeconomic assumptions for GDP, population growth, and fuel prices that are in line with Mexican government estimates made at the beginning of 2008. Under the baseline scenario, total CO_2e emissions are estimated to grow from 659 Mt in 2008 to 1,137 Mt in 2030.

Implementing the 40 MEDEC interventions that meet the criteria outlined for inclusion would reduce CO_2e by about 477 Mt in 2030 relative to the baseline (figure 1). Adopting these interventions would yield a level of emissions that is virtually the same as that in 2008, despite significantly higher GDP and per capita income. The emission reductions would come from agriculture and forestry (162 Mt), transport (131 Mt), electric power (91 Mt), energy end-use (63 Mt), and oil and gas (30 Mt). The emissions reduction potential of the MEDEC low-carbon scenario is conservative, in that only 40 interventions were considered and the analysis did not assume any major changes in technology.

How much would low-carbon development cost in Mexico, and how do the costs of interventions compare across sectors? Nearly half of potential emissions reduction comes from interventions that have positive net ben-

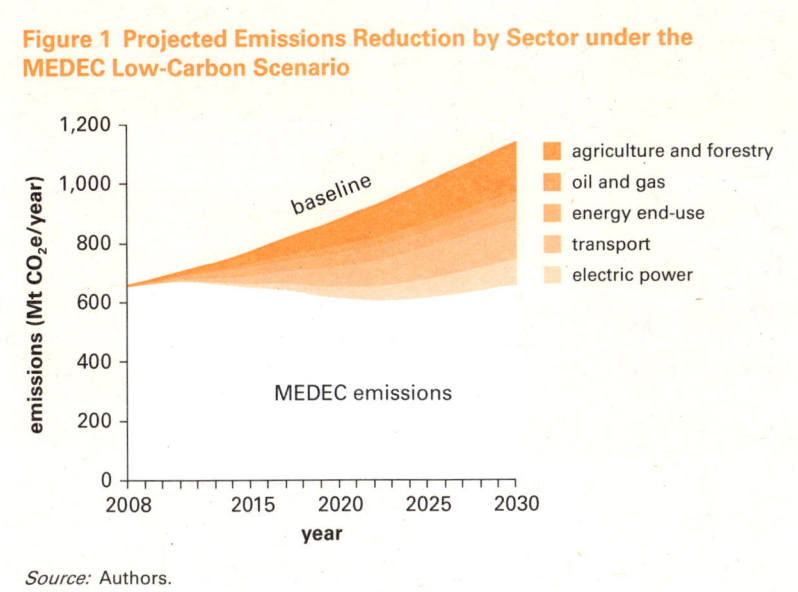

Figure 1 Projected Emissions Reduction by Sector under the MEDEC Low-Carbon Scenario

Source: Authors.

efits (negative costs), meaning that their overall cost is less than the respective high-carbon alternative (figure 2). Interventions that have both high potential and low cost include the following:

- Public transport and vehicle efficiency
- Most energy-efficiency measures, including electricity supply improvements, lighting, refrigeration, air conditioning, and improved cookstoves
- A number of low-cost energy supply options, including industrial (and Pemex) cogeneration and solar water heating

At a value of $10/t CO_2e, a number of other large interventions, including reforestation and restoration, and afforestation, yield positive benefits. Fully 80 percent of the greenhouse gas reduction potential of the MEDEC interventions lie below the $10/t CO_2e level. Raising the cost threshold to $25/t CO_2e allows more than 5 billion tons of CO_2e to be avoided through 2030.

Elements of a Low-Carbon Program

Many high-priority interventions in the transport, electric power, energy efficiency, and forestry sectors have net costs that are low or negative. The fact that many of these interventions have not already been adopted on a large scale suggests that there are barriers to implementing them.

Policies and Investments Required for Low-Carbon Development

Two of the greatest challenges Mexico will face in moving to a low-carbon economy are financing the (generally higher) upfront costs of low-carbon

Figure 2 Marginal Abatement Cost Curve

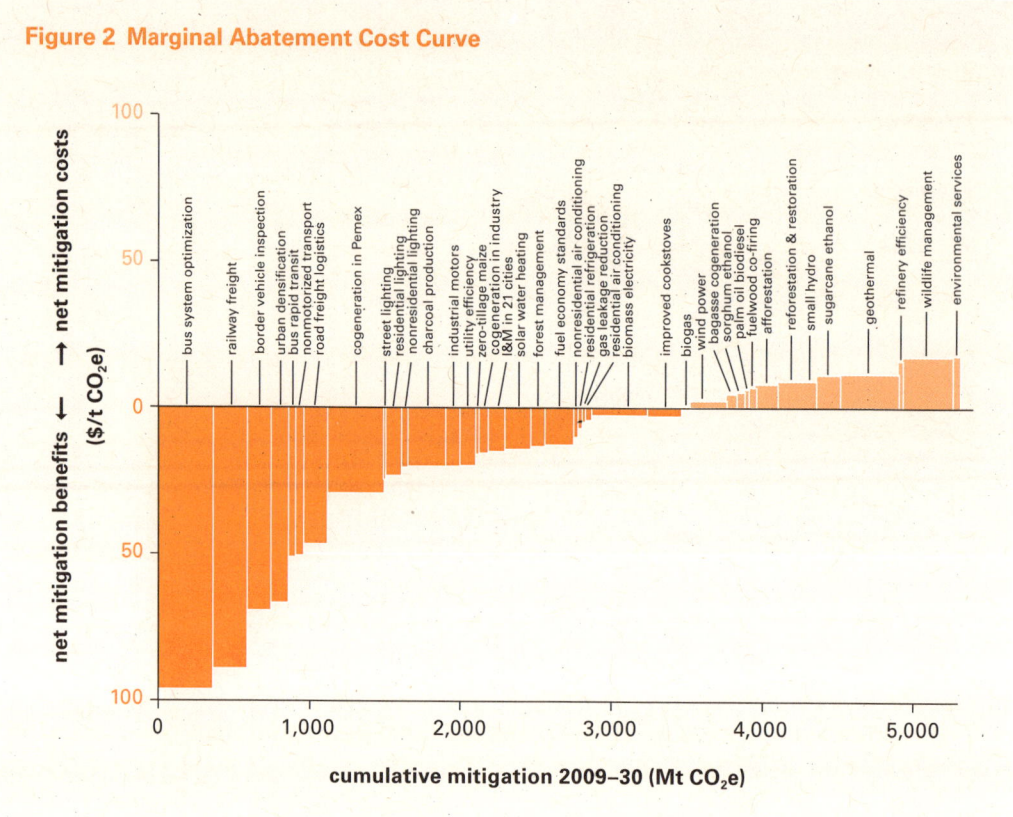

Source: Authors, based on MEDEC study results.

investments and putting in place supportive policies and programs to overcome the regulatory, institutional, and market development barriers. Renewable energy investments generally have higher initial costs than other investments. These costs are often compensated for by lower operating costs, yielding a net economic benefit (in present value terms). Even where the discounted life-cycle costs are lower, however, higher upfront investment costs often inhibit such investments. For some interventions, in particular in energy efficiency, the initial investments are offset by the savings in new generating capacity, resulting in "negative" investment cost differences when upstream effects are considered. The overall new investment required to achieve the MEDEC low-carbon scenario is about $64 billion between 2009 and 2030, or about $3 billion a year, equivalent to about 0.4 percent of Mexico's GDP in 2008.

Investment by the public sector will be critical, but financing will not have to come entirely from the government; there is considerable room to involve the private sector in financing investments in energy efficiency, renewable energy, and sustainable transport. The recent reform of the oil and gas industry represents a positive step in promoting greater efficiency in the sector and attracting investments from the private sector. Since the mid-1990s, there has been a dramatic increase in the number of independent power producers for natural gas power plants. This model could be improved and extended to promote investments in energy efficiency, cogeneration, and renewable energy generation.

Changing the rules that limit Pemex from tapping its cogeneration potential and providing substantial electricity production to the grid is a high priority for low-carbon development. Other important policies could include increasing energy-efficiency standards for both new and used vehicles; revising residential electricity tariffs and increasing the prices of petroleum products and natural gas; changing public procurement rules to facilitate investments in energy efficiency in schools, hospitals, government buildings, and municipal services; improving coordination by federal, state, and municipal governments and by different sector agencies at all levels of government concerning urban land-use planning and public transport; improving fuel quality and enforcing air quality standards; and expanding forest management programs.

Almost all of the MEDEC interventions have already been implemented in Mexico as commercial-scale investments projects or pilot programs, thus demonstrating the feasibility of implementing them in the near term. For many of the interventions, it is the scale-up from an individual project scale to a wider program that is needed. Scaling up these projects will require new policies and the financing of incremental investments, as well as other institutional and behavioral changes.

Some of the MEDEC interventions could be supported by resources from the Clean Development Mechanism (CDM) or other international carbon finance mechanisms. Most, however, would require new rules—in the context of either a reformed CDM or new mechanisms—to qualify for support. Understanding the mitigation potential, net costs, and implementation barriers is therefore crucial in the light of ongoing international climate negotiations.

Near-Term Priorities

Several low-carbon interventions could be implemented in Mexico in the near term. High-priority actions that have already been proven in Mexico and could be scaled up over the next five years include the following:

- Bus rapid transit, based on projects in Mexico and pioneered in other parts of Latin America
- Expansion of the efficient lighting and appliances programs developed by Fideicomiso para el Ahorro de Energía Eléctrica (FIDE) (Fund for Electricity Savings) and the Secretaría de Energía (SENER) (Ministry of Energy)
- Wind farm development in Oaxaca and elsewhere, based on CFE's pilots
- Avoided deforestation, based on the Los Tuxtlas project in Veracruz
- Cogeneration in Pemex facilities, based on the project at Nuevo Pemex.

Wherever Mexico's low-carbon development projects begin, there will be a need to experiment and gain experience, especially with new investment mechanisms and regulatory policies. To establish domestic support for a low-carbon program, Mexico should begin with measures that have positive economic rates of return. As the analysis shows, such interventions are plentiful. A second priority is to promote interventions that have positive social and environmental benefits, such as those with positive environmental externalities in the forestry sector and those that reduce local air pollution and health impacts in both sustainable transport and rural fuel use.

CHAPTER 1

Introduction

On May 25, 2007, President Felipe Calderón announced Mexico's National Climate Change Strategy (Estrategia Nacional de Cambio Climático [ENACC]), which put climate change at the center of Mexico's national development policy (SEMARNAT 2007). The ENACC established an initial blueprint for the long-term climate change agenda for the country, together with medium- to long-term goals for adaptation and mitigation. On August 28, 2009, Mexico published the PECC, which defines how to operationalize the ENACC during the coming three years, in particular by identifying priorities and financing sources, both domestic and international (PECC 2009). Like all government programs, the PECC is considered part of the 2007–12 Plan Nacional de Desarrollo (PND) (National Development Plan) and an integral part of the environmental sustainability pillar of the PND.[1] This study was conceived and has been carried out with the objective of contributing to Mexico's agenda on climate change mitigation.

Objectives of the Study

This study seeks to identify and evaluate low-cost options for reducing greenhouse gas emissions that Mexico can implement in the short to medium term. Specific objectives include the following:

- Evaluate low-carbon interventions by key sectors in Mexico using a common cost-benefit methodology (box 1.1), and identify barriers to implementing the interventions;
- Build a low-carbon scenario for Mexico to the year 2030 based on the potential and costs of the sectoral low-carbon interventions;
- Identify priority policies that would support a low-carbon development pathway, including a portfolio of low-carbon interventions that can be implemented now and within the next 5–10 years.

Box 1.1 Cost-Benefit Analysis Methodology

The economic analysis of low-carbon interventions uses a standardized cost-effectiveness frame-work for all sectoral interventions. The methodology is not technically a cost-benefit analysis, because it does not measure the benefits of climate change mitigation in terms of the reduction in climate change impacts but instead compares the costs of different interventions to reduce greenhouse gas emissions. In other words, the economic analysis does not assume a value for carbon mitigation but rather produces a "cost of carbon" as an output. The analysis calculates the net present value as of 2008 of the direct economic costs and benefits of each intervention between 2009 and 2030 to arrive at the "net costs" of reducing emissions.

The cost-effectiveness of reducing greenhouse gas emissions is thus the present value of the net cost of reducing (avoiding) 1 ton of CO_2 equivalent emissions. For each intervention over the study period, annual emissions reductions were added up to the cumulative emissions reduction; the stream of annual net costs was then discounted at 10 percent a year to determine the present value of the net cost in 2008. In the analysis, carbon was not discounted.

The net cost of the mitigation intervention is calculated by subtracting the direct benefits from the direct costs of implementing the intervention. The financial costs are reflective of the economic (social) opportunity costs to the extent that corrections were made for taxes and subsidies and traded goods were evaluated at their import and export parity values. Examples of direct benefits include energy cost savings or travel time and travel cost savings. Environmental externalities are considered as indirect co-benefits and are not included in the first-order cost-effectiveness calcula-tion shown in the marginal abatement cost curve. However, for some interventions in which health benefits from reduced air pollution are particularly important and damage functions have been estimated—such as transport and household fuelwood use—externality values were calculated (these results are discussed in the sector chapters, in box 5.1, and in chapter 7).

In the analysis of individual interventions, comparisons are made between the intervention and the baseline—the alternative that would have been pursued in the absence of the MEDEC program. Incremental net costs (benefits) are calculated by subtracting the costs (benefits) of the option from the costs (benefits) of the baseline case; and incremental net greenhouse gas emissions are calculated by subtracting the greenhouse gas emissions of the option from the greenhouse gas emissions of the baseline case. (For a more detailed explanation of the cost-benefit analysis, see appendix B.)

MEDEC builds on the ENACC and on the low-carbon development work program outlined in Mexico's Third National Communication, with the intention of providing tools for assessing and prioritizing low-carbon interventions and policies in Mexico. The study evaluates a broad range of potential low-carbon activities, comparing the results with international experiences and identifying strategic and competitive advantages of low-carbon development for Mexico, including opportunities for greater access to the carbon market and other resources for climate change mitigation.

The analyses focus on strategic sectors or themes of importance to Mexico that were jointly identified by the World Bank and the government of Mexico, following consultations with government agencies, academic insti-tutions, and public and private stakeholders. The new research undertaken for the study was intended to cover areas in which information was not abundant and to avoid overlap with earlier studies and projects. The sector analyses cover five themes:

- Power generation, which includes the production of electricity by centralized or decentralized power plants
- The oil and gas industry, which includes oil and gas extraction, pipelines, and refineries
- Energy end-use, which includes energy efficiency in the manufacturing and construction industries and the residential, commercial, and public sectors
- Transport (the single largest emitter of carbon dioxide equivalent [CO_2e] in Mexico), which includes primarily road transport
- Agriculture and forestry, which covers crop and timber production, forest and other land-use management, and a broad range of biomass energy.

The study also undertakes economic and emissions modeling and scenario analysis, in order to provide a broad perspective of opportunities and achievable goals, from an international perspective. The modeling uses the outputs of each sector analysis and develops emission scenarios to 2030.

The study conducts cost-benefit analyses of specific low-carbon opportunities for reducing greenhouse gas emissions in each sector, the financial requirements for investment in the sector, and the issues related to implementing the low-carbon development portfolio. Climate change mitigation options (referred to as "interventions") were selected based on their potential for greenhouse gas reduction, net costs (benefits), and feasibility in terms of political, social, institutional, legal, and other preconditions. The interventions identified are presented both by sector and individually, in order to allow the government or other institutions to assess what a combination of reduction activities would entail in terms of investment costs and reduction potential and to be able to assess this flexibly within the framework of political conditions, available resources, and other considerations.

Strategic Significance to Mexico of Low-Carbon Development

Mexico could benefit from moving to a low-carbon economy for at least four reasons:

- It is likely to suffer disproportionately from the impacts of climate change and therefore has a strong interest in ensuring that an international agreement to limit emissions is adopted.
- Various "no-regrets" interventions (that is, interventions that have positive economic rates of return and should be undertaken regardless of climate change considerations) can contribute substantially to the country's economic development.

- Many interventions yield important co-benefits, such as energy security, human health benefits, and environmental protection.
- Countries that pursue low-carbon development are likely to enjoy strategic and competitive advantages.

Mexico faces high risks from climate change with respect to water availability, the increased frequency and intensity of tropical storms, and potential inundation from two ocean coastlines. Initially, the impact of climate change was expected to be felt only over the longer term; there is now increasing evidence that climate change impacts are already occurring.

The Fourth Assessment Report of the Intergovernmental Panel on Climate Change (IPCC) predicts that under baseline scenarios, relative to 1961–90, temperature increases in Latin America and the Caribbean could reach 0.4°–1.8°C by 2020 and 1°–4°C by 2050 (De la Torre, Fajnzylber, and Nash 2009). These projections, derived from global circulation models, also forecast changing precipitation patterns across the region (Christensen and others 2007). The predictions of at least five of eight global climate models indicate that by 2030 the number of consecutive dry days in Mexico will increase and heat waves will become longer. Midrange climate forecasts indicate that arid regions of the country are likely to experience severe species loss by 2050, losing 8–26 percent of their mammal species, 5–8 percent of their bird species, and 7–19 percent of their butterfly species (De la Torre, Fajnzylber, and Nash 2009).

Damage to the Gulf Coast wetlands in Mexico is a serious concern. Global circulation models agree that the Gulf of Mexico is most vulnerable to the impacts of climate change of all coastal areas in the region; Mexico's three national communications to the United Nations Framework Convention on Climate Change (UNFCCC)—in 2001, 2004, and 2007—document ongoing damage, raising urgent concerns about the area's security from climate change. Wetlands in this region are currently suffering from man-made impacts associated with land-use changes, mangrove destruction, pollution, and water diversion, which make the ecosystem even more vulnerable to the impacts of climate change. The total mangrove area in the Gulf Coast region is disappearing at an annual rate of 1 to 2.5 percent. As a result of climate change, Mexico may experience 10–20 percent decreases in water runoff nationally and as much as a 40 percent decline over the Gulf Coast wetlands. These wetlands possess the most productive ecosystem in the country and one of the richest on Earth (Vergara 2008).[2] About 45 percent of Mexico's shrimp production, for example, originates in the Gulf wetlands, as does 90 percent of the country's oyster crop and at least 40 percent of commercial fishing volume.

Data also suggest a trend toward storms and weather-related natural disasters in Mexico and surrounding countries that are more frequent, stronger, or both. Extreme weather events already exact a high toll in the region. In 1998 Hurricane Mitch killed at least 11,000 and perhaps as many as 19,000 people across Central America and Mexico. In 2005 Hurricane Wilma, the strongest Atlantic hurricane on record, damaged 98 percent of the infrastructure along the northeastern coast of Mexico's Yucatan Peninsula, home to Cancun, and inflicted an estimated $1.5 billion loss on the tourism industry.

Mexico hopes it can benefit strategically and economically by moving to a low-carbon economy and tapping local opportunities and advantages. Many policies and actions it can take to reduce greenhouse gas emissions can improve energy security, enhance the country's competitive position and trade balance, and reduce local environmental damage.

Previous studies have identified several promising areas for mitigating climate change in Mexico:

- Expanding energy efficiency and the development and use of renewable energy
- Increasing domestic gas production, and improving the overall efficiency of the sector (such as reducing gas losses), in order to meet the country's growing demand for natural gas, improve local air quality, increase energy efficiency in power and industry, and reduce the growing dependence on imports of gas from the United States
- Avoiding deforestation and implementing reforestation and afforestation projects, which can reduce greenhouse gas emissions in Mexico while contributing to biodiversity preservation, water and soil management, and improved local livelihoods.

The benefits to Mexico of taking a stronger position on climate change and promoting low-carbon development are competitive and strategic. The federal government, which has taken a proactive position on climate change, recognizes these benefits.

Greenhouse Gas Emissions in Mexico

Mexico emitted 643 million tons of carbon dioxide equivalent (Mt CO_2e) in 2002 (Third National Communication to the UNFCCC). About 390 Mt CO_2e—61 percent of total emissions—was generated from fossil fuel–based energy production and consumption, including significant fugitive emissions (leakage, venting, flaring) in oil and gas production and transportation. The remaining emissions were from land use, land-use change, and forestry (LULUCF) (14 percent); waste (10 percent); industrial processes (8 percent); and agriculture and livestock (7 percent).

Mexico ranks 13th in the world in total greenhouse gas emissions and is the second largest emitter in Latin America after Brazil. Mexico accounts for 1.4 percent of global CO_2e emissions from energy consumption; it is the largest emitter in Latin America if land-use change and forestry emissions are excluded. Mexico's CO_2e emissions from energy consumption are greater than those of Brazil and South Africa but significantly below those of China or India. Its total greenhouse gas emissions are equivalent to about 6 t CO_2e per capita; or about 4 t CO_2e per capita if only emissions from fossil fuel combustion are included (figure 1.1).

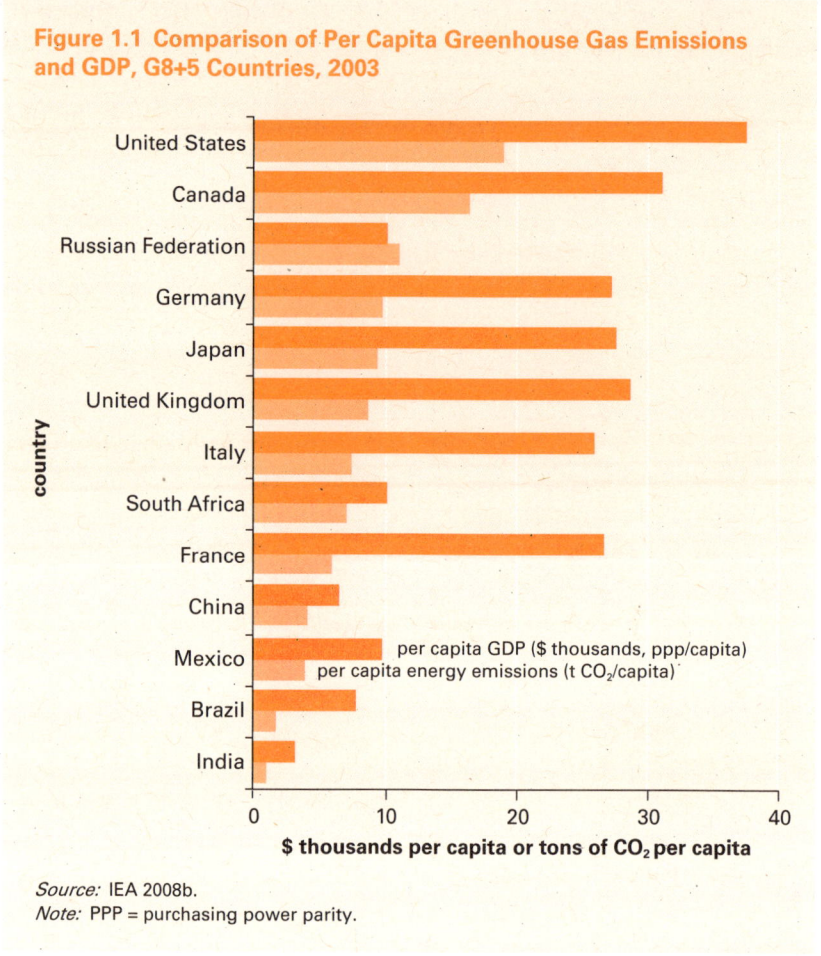

Figure 1.1 Comparison of Per Capita Greenhouse Gas Emissions and GDP, G8+5 Countries, 2003

per capita GDP ($ thousands, ppp/capita)
per capita energy emissions (t CO_2/capita)

Source: IEA 2008b.
Note: PPP = purchasing power parity.

Excluding LULUCF, for which emissions estimates are less certain than they are for energy consumption, Mexico's greenhouse gas emissions increased 30 percent from 1990 to 2002 (figure 1.2). Emissions from waste experienced the fastest growth, almost doubling in quantity, driven by increased solid waste and wastewater. Emissions from industrial processes also grew significantly, in large part because of booming construction in this period, which increased the use of limestone and dolomite as well as the production of building materials, such as cement, iron, and steel. Agricultural emissions, which include emissions from livestock, fertilizers, and soil carbon, declined about 3 percent during the same period, mainly as a result of the decline in fertilizer use.

Greenhouse gas emissions from energy production and consumption activities grew steadily between 1990 and 2002, accounting for 60 percent of the overall increase in emissions (figure 1.3). Increased fossil fuel consumption in power generation and transport accounted for about 90 percent of the increment in greenhouse gases associated with energy production and consumption.

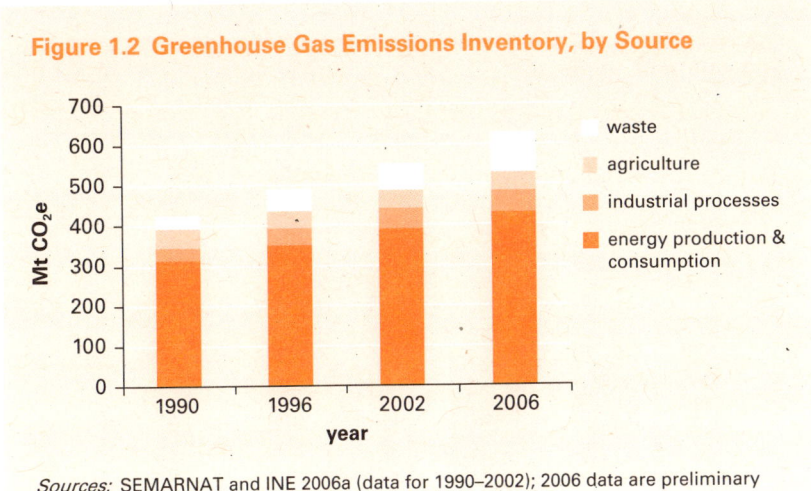

Figure 1.2 Greenhouse Gas Emissions Inventory, by Source

Sources: SEMARNAT and INE 2006a (data for 1990–2002); 2006 data are preliminary and are from INE.
Note: Data exclude emissions related to land use, land-use change, and forestry.

LULUCF is an important source of Mexico's greenhouse gas emissions. Estimates based on new information put the net emission of LULUCF at about 103 Mt CO_2e in 2005, a sizable increase over the 90 Mt CO_2e figure in the 2002 national inventory. Over the longer term, with improved forestry management and an overall balance between deforestation and reforestation or afforestation, LULUCF could become a net sink of greenhouse gases in Mexico.[3]

Mexico's Climate Change Actions

Recognizing the threat climate change poses to its development, Mexico has been among the most active countries in international climate change discussions. As a non–Annex I country,[4] Mexico is not mandated to limit or reduce its greenhouse gas emissions under the Kyoto Protocol, but the country has firmly adopted the UNFCCC principle of "common but differentiated responsibilities" and pledged to reduce its emissions on a voluntary basis.

Mexico has submitted three National Communications to the UNFCCC. The First National Communication (1997) established the national greenhouse gas inventory and reported the first studies on Mexico's vulnerability to climate change. The Second National Communication (2001) updated the national greenhouse gas inventory to cover 1994–98 and included future emission scenarios. The Third National Communication (2006) updated the national greenhouse gas inventory to 2002 and included land-use change emissions estimates for 1993–2002 and a number of mitigation and adaptation studies (SEMARNAP and INE 1997; SEMARNAT and INE 2001, 2006b). Mexico is the only non–Annex I country to have submitted a Third National Communication and is currently preparing its Fourth National Communication.

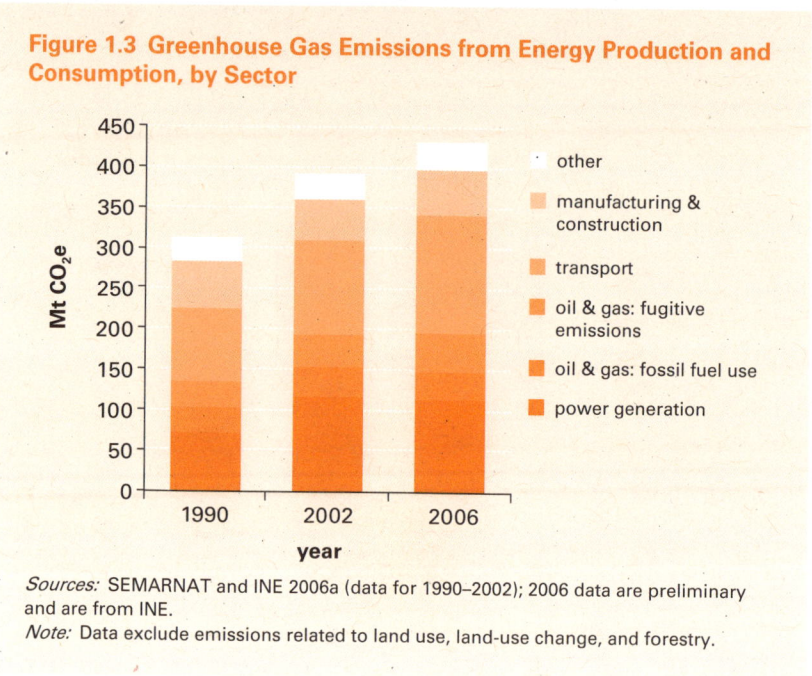

Figure 1.3 Greenhouse Gas Emissions from Energy Production and Consumption, by Sector

Sources: SEMARNAT and INE 2006a (data for 1990–2002); 2006 data are preliminary and are from INE.
Note: Data exclude emissions related to land use, land-use change, and forestry.

Recognizing the multisectoral challenges posed by climate change, in April 2005 Mexico established the Comisión Intersecretarial de Cambio Climático (CICC) (Intersecretarial Commission on Climate Change). The CICC's key mandates include formulating and coordinating national climate change strategies and incorporating them into sectoral programs.[5] The CICC contains several working groups, including groups on mitigation and adaptation. Associated with the CICC is an advisory board on climate change, which creates a link between the CICC, the scientific community, and civil society (see http://tinyurl.com/infoc4).

Overview of the Sector Analysis and Structure of the Report

Chapters 2–6 assess the potential for greenhouse gas reduction in Mexico by sector. For the purposes of analysis, the economy was divided into five primary sectors: electric power; oil and gas; stationary energy end-use sectors (including residential, industrial, commercial, and service sectors); transport; and agriculture and forestry (including biomass energy). These sectors, chosen based on their importance to current and projected future emissions, cover more than 90 percent of Mexico's current emissions.[6] The sectoral work draws on detailed background reports prepared for MEDEC.

Each sectoral analysis focuses on a set of MEDEC interventions that would reduce greenhouse gas emissions over the coming two decades. The emission reduction interventions were selected based on their potential for overall emissions reduction, the net costs of interventions that reduce emissions, and the feasibility of implementing the interventions in the short to medium term (box 1.2).

Forty interventions were selected (table 1.1). Many are cross-sectoral or occur in one sector but have effects in others. In particular, several interventions in the industrial, oil and gas, and agriculture and forestry sectors generate electricity and thus mitigate greenhouse gas production in the electricity sector. Most energy end-use efficiency interventions reduce electricity consumption.

The majority of interventions in the agriculture and forestry sector reduce greenhouse gas emissions through "avoided deforestation" and by actively building up carbon stocks in woody biomass and in soils. Other agricultural interventions include the substitution of fossil fuels by liquid biofuels, reducing emissions in the transport sector. Some forestry interventions have multiple impacts: they produce biomass energy that substitutes for fossil fuel use in other sectors, and they contribute to a reduction in deforestation and forest degradation.

Chapters 2–6 present the findings of the detailed sectoral work undertaken as part of MEDEC. The results of the analysis of the low-carbon interventions for each sector are aggregated in chapter 7 to form a scenario for low-carbon development in Mexico through 2030. The relative costs of the interventions are compared in chapter 7 in the form of a marginal abatement cost curve. Chapter 8 discusses the conclusions of the low-carbon scenario analysis in terms of the feasibility of implementing a program of interventions and a portfolio of projects that could be carried out in the near term.

Box 1.2 Criteria for Selecting Interventions

Three principal criteria have been used to identify low-carbon interventions for analysis in the MEDEC study: the potential for reducing greenhouse gas emissions, the net cost of doing so, and the feasibility of implementation.

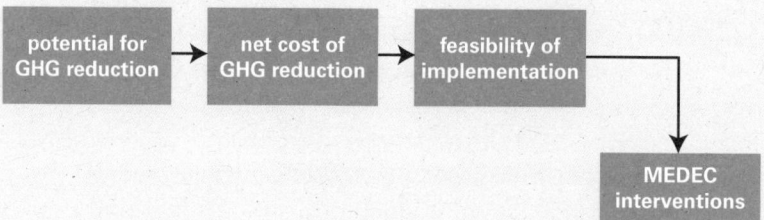

The first criterion is that low-carbon interventions should have substantial potential for reducing greenhouse gas emissions. For the purposes of this study, 5 million tons (Mt) of CO_2e emissions reduction implemented between 2009 and 2030 was used as the threshold for including an intervention. Some interventions that did not meet the 5 Mt CO_2e threshold may have excellent economic and social returns and should be pursued under domestic or international carbon programs. (For example, evaluation of the collection and use of animal waste determined that this intervention did not meet the threshold reduction target. A number of animal biogas projects are being undertaken in Mexico, several with carbon revenues. Such projects may be excellent candidates for support under a climate mitigation program.) Such interventions were not included in this study.

The second criterion is that low-carbon interventions should be low cost. Interventions should have positive economic and social rates of return (at a given discount rate or cost of capital). Many interventions have positive net benefits. In these cases CO_2 reduction is free, because the other financial and economic benefits of the intervention more than cover the costs. Such projects are often referred to as "no-regrets" projects, because society should be undertaking them even in the absence of climate change considerations. Other interventions have net costs. In these cases the cost per ton of CO_2e should be low. An upper bound of $25 per ton CO_2e was used for selecting these interventions.

The third criterion is that low-carbon interventions should be feasible in the short or medium term. This criterion is the most challenging and requires discussions with sectoral experts, government officials, the private sector, and civil society. For the purposes of selecting the MEDEC interventions, "feasibility" was first determined by sectoral experts in terms of their technical potential, market development, and institutional requirements. (The MEDEC interventions assumed reliance on existing technologies; any productivity gains and related cost reductions would be caused largely by changes in the scale of production.) Most of the selected interventions were also discussed with government officials, to assess the political and institutional feasibility of expanding the intervention in Mexico. (All MEDEC interventions have already been implemented, at least on a pilot level, in Mexico or in other countries with similar conditions. Some interventions face barriers in the short term, but it was believed that these barriers can be removed in the medium term.) Finally, the interventions were subjected to a review by World Bank staff to ensure that the measures were feasible in a broader context, both from a market perspective and with respect to sustainability criteria, such as environmental and social safeguards. (A discussion of the social, political, institutional, and financial barriers to low-carbon interventions and the policies that could be used to overcome them is provided in the sector and final chapters.)

Table 1.1 MEDEC Interventions by Sector

Sector	Intervention	Emissions reduction				
		Electricity	Heat	Transport	Land use	Other[a]
Electric power	Wind power	■				
	Geothermal power	■				
	Small hydropower	■				
	Biogas	■				
	Utility efficiency	■				
Oil and gas	Cogeneration in Pemex	■	■			
	Refinery efficiency		■			
	Gas leakage reduction					■
Energy end-use	Bagasse cogeneration	■	■			
	Cogeneration in industry	■	■			
	Residential air conditioning	■				
	Residential lighting	■				
	Street lighting	■				
	Industrial motors	■				
	Nonresidential lighting	■				
	Nonresidential air conditioning	■				
	Residential refrigeration	■				
	Solar water heating		■			
	Improved cookstoves		■			
Transport	Urban densification			■		
	Bus rapid transit systems			■		
	Nonmotorized transport			■		
	Bus system optimization			■		
	Vehicle fuel economy standards			■		
	I&M in 21 cities			■		
	Border vehicle inspection			■		
	Road freight logistics			■		
	Railway freight			■		
Agriculture and forestry	Biomass electricity	■			■	
	Fuelwood co-firing retrofitting	■			■	
	Charcoal production		■		■	
	Zero-tillage maize			■	■	
	Reforestation and restoration				■	
	Afforestation				■	
	Wildlife management				■	
	Forest management				■	
	Payment for environmental services				■	
	Palm oil biodiesel			■		
	Sorghum ethanol			■		
	Sugarcane ethanol	■		■		

Source: Authors.
Note: I&M = inspection and maintenance.
a. "Other" includes industrial processes, waste, flaring, and fugitive emissions.

Notes

1. The main objective in this pillar is to turn the concept of environmental sustainability into a cross-cutting element of public policies and ensure that all public and private investments are compatible with environmental protection. Objectives and strategies are structured in such areas as water, forests, climate change, biodiversity, solid waste, and cross-sectoral environmental sustainability policy instruments.

2. The Mexican National Institute of Ecology (INE) has identified wetlands in the Gulf of Mexico as one of the ecosystems most threatened by anticipated climate changes (data published on projected forced hydroclimatic changes, as part of IPCC assessments [Vergara 2008]). This has been documented in Mexico's third national communication to the UNFCCC.

3. The uptake and storage of carbon by plants and soil is often referred to as a "sink" of CO_2 from the atmosphere. Thus, for example, if the amount of carbon absorbed by forests is greater than the CO_2 emissions from forests, such as through forest fires or soil degradation, there is said to be a net sink of CO_2.

4. Annex I countries are signatories to the UNFCCC (and the Kyoto Protocol) that agree to reduce their greenhouse gas emissions to targets set by the Convention. Non–Annex I countries include developing countries and economies in transition that do not have mandatory reduction targets under the Kyoto Protocol.

5. The CICC is chaired by the Minister of Environment and Natural Resources (SEMARNAT), the Vice-Minister of Environment Planning serves as Executive Secretary with Ministers of the following areas serving as members: the Ministry of Agriculture, Livestock Production, Rural Development, Fisheries and Food (SAGARPA), the Ministry of Communication and Transportation (SCT), the Ministry of Economy (SE), the Ministry of Social Development (SEDESOL), the Ministry of Energy (SENER), and the Ministry of Foreign Affairs (SRE). The Ministry of Finance (SHCP) is a permanent invited member to the CICC's deliberations. For more details, see http://tinyurl.com/infocicc.

6. The sectors that are not covered by MEDEC are waste and industrial processes. Some relevant mitigation opportunities exist in these sectors. In particular, wastewater treatment plants and landfill sites have significant potential for capturing and burning methane or for using it for energy purposes.

CHAPTER 2

Electric Power

Mexico's electric power sector is the second-largest greenhouse gas emitter after transport, accounting for about 26 percent of greenhouse gas emissions from energy production and consumption (see figure 1.3). Electricity production is expected to grow significantly in Mexico over the coming decades to meet the needs of an expanding economy and growing population. The technologies and fuel mix for power generation will have a major impact on the resulting greenhouse gas emissions from the sector.

The Mexican electricity system is dominated by two state-owned companies—Comisión Federal de Electricidad (CFE) and Luz y Fuerza del Centro (LyFC)[1]—which handle generation, transmission, and distribution of electricity and serve more than 97 percent of the population. CFE provides service to most of the country outside the capital; LyFC operates in Mexico City and surrounding areas.

Since the late-1990s, new generating capacity has been provided primarily by independent power producers (IPPs) that generate and sell power exclusively to CFE under long-term contracts. In 2007 IPPs represented about 23 percent of total installed capacity in Mexico and generated 31 percent of total electricity. As of 2007, the total installed capacity of the electric power system, including self-supply and export projects, was 59,209 MW, which generated 262 TWh a year.

About 76 percent of Mexico's installed generation capacity is fired by fossil fuels—fuel oil, natural gas, coal, and small amounts of diesel. The remaining capacity consists of hydropower (19 percent), nuclear (2.3 percent), geothermal (1.6 percent), bagasse (sugarcane pulp) and other biomass (0.6 percent), and a small fraction of wind power.

The most notable change in the generation mix over the past decade has been the large increase in natural gas–fired plants, which have replaced fuel oil plants. The use of natural gas in the power sector increased at an average annual rate of about 16 percent between 1997 and 2007, reaching an installed capacity of about 20,000 MW (excluding self-supply). Natural gas consumption by the power sector reached 27,300 Mm³ in 2007, equivalent to 38 percent of total domestic gas consumption (SENER 2008b). Coal-fired plants entered the mix in the early 1980s and have gradually increased to 7.9 percent of installed capacity. Despite public and regulatory pressure to reduce coal use in some industrial and middle-income countries based on environmental considerations, the overall international trend, driven by investment and fuel costs, is toward further expansion of coal-fired capacity. Mexico's hydropower capacity increased by 50 percent in absolute terms over the past two decades, but its share in total capacity fell from 30 percent to 19 percent. The large share of gas-fired generation and sizable portion of hydropower contributed to the relatively low carbon intensity of electricity in Mexico relative to most G8+5 countries (figure 2.1).

In 1997 the government of Mexico created a financial mechanism—Proyectos de Impacto Diferido en el Registro de Gasto (Projects with Differed Expenditure Impact) (PIDIREGAS)—to finance long-term oil, gas, and power projects with government-guaranteed private investment. Under this scheme and through traditional budget financing, CFE increased installed capacity by more than 15 GW between 1999 and 2008, including

Figure 2.1 Electric Power Generation by Fuel Type in Selected Countries, 2005

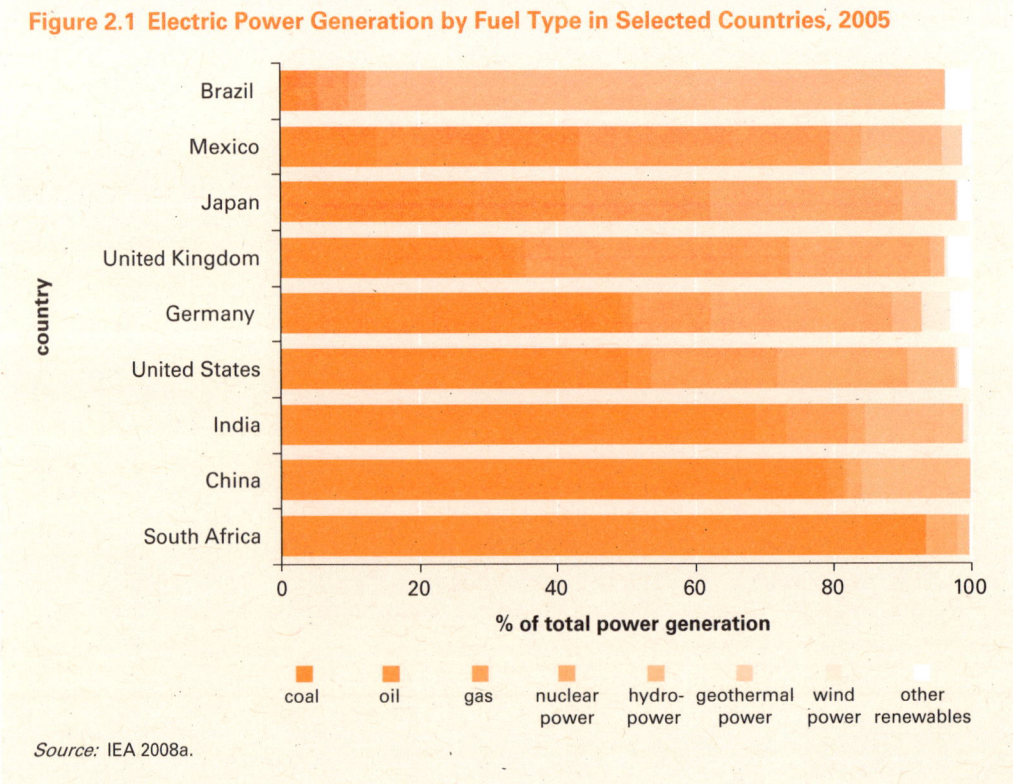

Source: IEA 2008a.

11 GW through IPP contracts based on combined-cycle natural gas plants. As of 2008, Mexico had a surplus of generating capacity, with an operating reserve margin of 21 percent (15 percent is standard).

Technical transmission losses have been declining in percentage terms in both CFE and LyFC, partly as a result of an ambitious investment program in CFE, financed through the PIDIREGAS scheme. As of 2005, technical transmission losses were less than 2 percent for CFE, which is on par with good international practice, and 3 percent for LyFC (Komives and others 2009).

In contrast, distribution losses in both companies are high by international standards, and they have been increasing in recent years. CFE's technical and commercial distribution losses rose from 11.0 percent in 2000 to 11.6 percent in 2005. (Good international practice would be about 8 percent for a utility with CFE's load and geographic characteristics.) LyFC's distribution losses are very high, having exceeded 30 percent since 2005. Overall, technical and commercial losses of Mexico's electricity system represent 16.2 percent of electricity generation (figure 2.2).

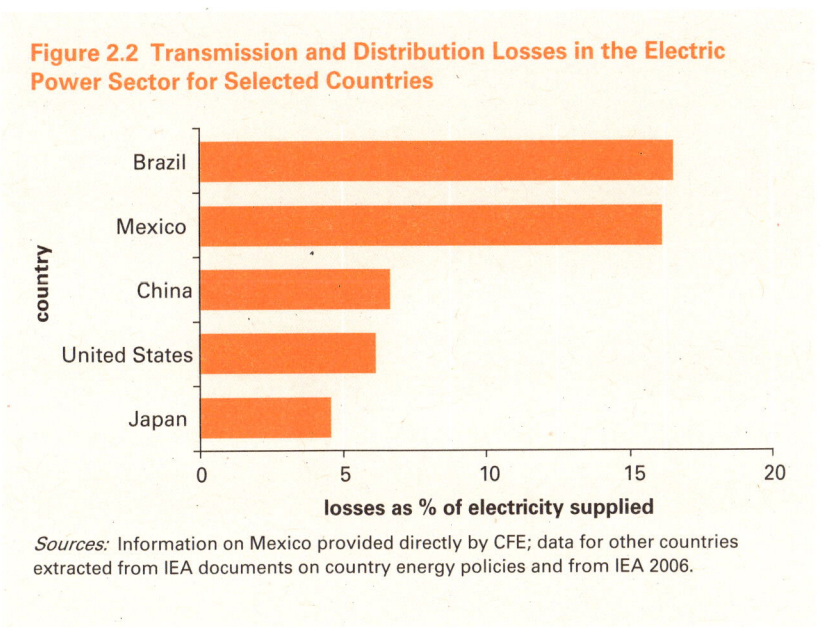

Figure 2.2 Transmission and Distribution Losses in the Electric Power Sector for Selected Countries

Sources: Information on Mexico provided directly by CFE; data for other countries extracted from IEA documents on country energy policies and from IEA 2006.

The Baseline Scenario

The government projects electricity demand to grow 4.8 percent a year between 2007 and 2016, compared with projected annual GDP growth rate of 3.0–3.5 percent.[1] This growth path follows the historical trend, in which electricity consumption has grown significantly faster than GDP. Meeting this rising demand will require the addition of 2,040 MW of new capacity each year on average. Annual average investments—for generation, transmission, distribution, and related fuel-handling facilities, such as ports and processing facilities—are estimated at about $5.5 billion.[2]

The baseline scenario uses the government's demand projections for the period to 2016. For the period 2017–30, it assumes that electricity generation increases 3.9 percent a year, reaching 630 TWh by 2030. Installed capacity (not including self-supply) is projected to increase by a factor of 2.2, from about 50 GW in 2008 to 110 GW in 2030.[3]

The selection of power-generation technologies for 2017–30 was based on the assumptions that expansion is based on demand projections and least-cost technology[4] and that environmental requirements for criteria pollutants (particulates, SO_2, and NO_X) are met. Unlike the government's current planning outlook, which sets a ceiling on coal penetration, the baseline scenario assumes that power-supply technologies are driven primarily by costs, without consideration of climate change or other policy-driven issues. The large increase in coal-based electricity generation under the baseline scenario is consistent with recent trends in a number of countries worldwide.

Under these assumptions, there would be a distinct shift in the fuel mix of Mexico's power sector by 2030, with a nearly 6-fold increase in coal-fired generation requiring significant investments in coal-related infrastructure and a 2.5–fold increase in gas-fired power generation (figure 2.3). Both coal and gas imports for power generation would rise significantly.

Figure 2.3 Electricity Generation by Fuel Type in Mexico: Historical Trend and Projected Growth under the Baseline Scenario, 1965–2030

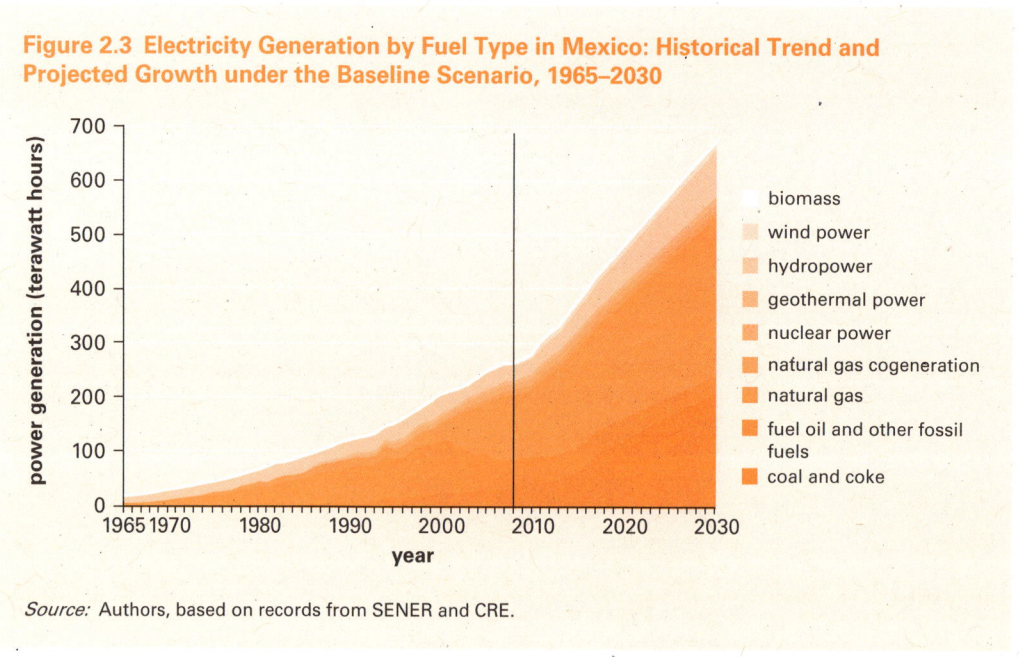

Source: Authors, based on records from SENER and CRE.

Under the baseline scenario, total CO_2e emissions from power generation increase 230 percent, from 142 Mt CO_2e in 2008 to 322 Mt CO_2e in 2030 (figure 2.4). The expansion of coal-fired generation accounts for 33.5 percent of the increase; gas-fired generation accounts for 46.2 percent. Despite the much larger share of coal-fired generation, the overall carbon intensity of electricity production drops in the baseline, from 0.538 t CO_2e/TWh in

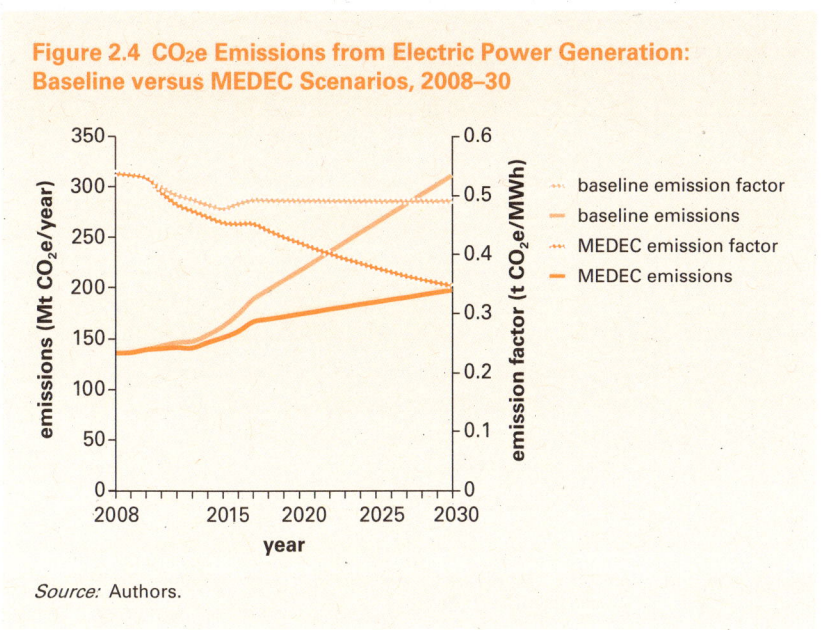

Figure 2.4 CO_2e Emissions from Electric Power Generation: Baseline versus MEDEC Scenarios, 2008–30

Source: Authors.

2008 to 0.493 t CO_2e/TWh in 2030, because of the larger contribution of hydropower and natural gas and the smaller contribution of fuel oil.

The MEDEC Low-Carbon Scenario

Under the MEDEC low-carbon scenario, reduction of greenhouse gas emissions is introduced as an explicit goal of power-capacity expansion. No attempt is made to reoptimize the power expansion plan of the baseline scenario by imposing an arbitrary greenhouse gas mitigation constraint. Instead, a range of power supply options and technologies is evaluated. As the baseline already assumes a significant decrease in fuel-oil use, the low-carbon technologies are compared with the other two dominant power generation technologies in the baseline that contribute significantly to CO_2e emissions: natural gas power plants (combined-cycle technology) and coal-fired power plants (supercritical technology).

The MEDEC scenario is constructed by replacing new power capacity from these technologies under the baseline with suitable lower-carbon options and generation technologies (table 2.1). The potential for each low-carbon technology is assessed considering the availability of renewable resources in Mexico and the technical feasibility of integrating intermittent energy into the system. Based on international experiences of electricity systems with relatively large shares of intermittent energy sources, the reliability of the Mexican electricity system is not expected to be reduced by implementation of the MEDEC scenario. Furthermore, since the MEDEC scenario includes a mix of technologies offering base-load (geothermal), intermittent (wind), and peak generation (biomass,

Table 2.1 Levelized Costs of Main Power Generation Technologies
$/MWh

Technology	Generation investment	Exploration investment	O&M costs	Nonfossil fuel costs	Fossil fuel costs	Total
Baseline technologies						
Combined-cycle gas	19.57	n.a.	4.08	n.a.	55.17	78.98
Supercritical coal	30.97	n.a.	6.49	n.a.	18.33	55.79
Large hydropower	83.42	n.a.	1.55	3.58	n.a.	88.55
Gas turbine	68.88	n.a.	9.62	n.a.	82.12	160.62
MEDEC technologies						
Wind power	58.79	n.a.	10.45	n.a.	n.a.	69.24
Small hydropower	71.84	n.a.	13.50	3.58	n.a.	88.92
Geothermal power	40.18	31.52	24.23	n.a.	n.a.	95.92
Biogas	52.60	n.a.	10.29	n.a.	n.a.	62.88
Cogeneration in Pemex	40.50	n.a.	−$4.71	n.a.	−138.95	−103.16
Cogeneration in industry	25.18	n.a.	4.89	n.a.	39.10	69.17
Bagasse cogeneration	99.12	n.a.	n.a.	n.a.	−22.27	76.85
Biomass electricity	40.37	n.a.	18.33	−7.48	0.34	51.55

Sources: World Bank 2008; CFE 2008a.
Note: n.a. = not applicable; O&M = operations and maintenance. Exploration costs for fossil fuels are not included, because they are reflected in fossil fuel costs. Externalities are not included in the estimates.

and most small hydro and cogeneration), it is assumed that the demand for power can be met.

Investment costs for the different generation technologies are based on international references (World Bank 2006b, 2008), assuming no major changes in technology over the scenario period.[5] The operations and maintenance costs and fuel consumption figures reflect local conditions in Mexico (CFE 2008a). Fuel prices are based on common macroeconomic projections used in all sectors and reflect international trends. The cost analysis also estimates health damage costs, based on published valuations of externalities of SO_2, NO_X, and particulates (PM10), but they are not included in the marginal abatement cost assessment—nor are they included in table 2.1.

The MEDEC scenario assumes that generation technologies with a net cost below $25/t CO_2e will be deployed. Under this scenario, the share of coal declines significantly relative to the baseline scenario, from 31 percent to 6 percent, and the contribution of low-carbon technologies increases substantially (figure 2.5). The share of power generation increases from 2.0 percent to 11.0 percent for geothermal, from 0.1 percent to 8.0 for biomass, from 1.3 percent to 6.0 percent for wind, and from 0.4 percent to 2.5 percent for small hydro. Relative to the baseline, implementing the MEDEC scenario requires estimated net investment of $10 billion for the electric power sector.

Figure 2.5 Electric Power Generation by Fuel Type in Baseline versus MEDEC Scenarios

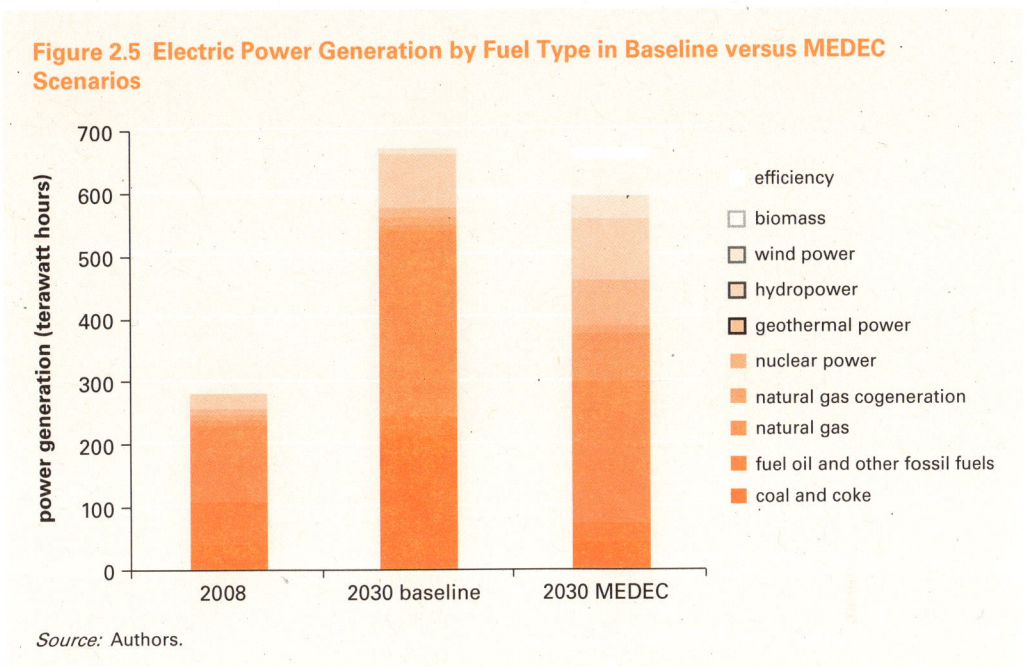

Source: Authors.

Five interventions are included (table 2.2). Four deploy renewable energy technologies for the generation of electricity (wind, small hydro, geothermal, and biogas). One entails energy-efficiency improvements in

Table 2.2 Summary of MEDEC Interventions in the Electric Power Sector

Intervention	Capacity (MW)	Maximum annual emissions reduction (Mt CO_2e/year)	Net cost or benefit of mitigation ($/t CO_2e)
Utility efficiency	n.a.	6.2	19.3 (benefit)
Electricity generation			
Biogas	940	5.4	0.6 (cost)
Wind power	10,800	23.0	2.6 (cost)
Small hydropower	2,750	8.8	9.4 (cost)
Geothermal power	7,500	48.0	11.7 (cost)
Electricity generation in other sectors[a]			
Cogeneration in Pemex	3,690	26.7	28.6 (benefit)
Cogeneration in industry	6,800	6.5	15.0 (benefit)
Bagasse cogeneration	2,000	6.0	4.9 (cost)
Biomass electricity	5,000	35.1	2.4 (benefit)
Fuelwood co-firing retrofitting	2,100	2.4	7.3 (cost)

Source: Authors.
Note: n.a. = not applicable.
a. See chapters 3, 4, and 6 for descriptions of electricity generation interventions in other sectors.

public utilities, including in transmission, distribution, and auxiliary equipment in existing power plants.[6] Several interventions in the electricity sector were considered and assessed but ultimately not included in the MEDEC scenario, because they did not meet the MEDEC criteria, because data were not available, or for other reasons. In particular, the generation of electricity from concentrated solar power or grid-connected photovoltaic technologies is set to become a relevant mitigation option in the coming decades, but mitigation costs are still well above the 25 $/tCO2e threshold. The generation of electricity from nuclear power in Mexico faces a series of security, environmental, and economic constraints. The rehabilitation of existing power plants, including thermal and hydro, is in many cases a cost-effective option, but was not analyzed owing to a lack of data.

Barriers to Mitigating Greenhouse Gas Emissions

The deployment of mitigation interventions in the electric power sector faces significant policy and institutional biases against two important low-carbon alternatives: cogeneration and renewable energy. Additional barriers to implementation are identified in table 2.3.

The power sector is designed to operate with current conventional, centralized generation technologies. Although in many cases cogeneration and renewable energy can compete with conventional technologies in Mexico in terms of cost, such technologies have scale and availability characteristics that are not conducive to centralized control. Utility procurement rules, for example, exclude in practice small-scale projects.

Current power generation planning methods do not account for important co-benefits offered by low-carbon technologies. In addition to climate mitigation, these benefits can include reducing local environmental and health impacts, increasing the security of the energy supply, diversifying the sources of energy and reducing risk, and enhancing industrial competitiveness by increasing efficiency.

There is significant potential to reduce greenhouse gas emissions through small hydropower generation at moderate incremental costs. Development of this source of energy is hindered, however, by relatively large capital costs and the high level of uncertainty over water concession licenses, which are provided by the Comisión Nacional del Agua (CONAGUA) (National Water Commission), and over the availability of water once the plant is in operation, when the resource will be shared with other uses, such as fishing and irrigation. The schedule for resource sharing is determined by CONAGUA, which has traditionally given priority to nonpower activities. This practice significantly increases the financial risk of hydropower projects and has discouraged private participation in small-scale hydro projects under the self-supply scheme. At the same time, many existing water supply and irrigation facilities could be equipped for electricity generation. Preliminary estimates suggest that more than 70 irrigation dams in Mexico could be used for power-generation purposes (CONAE 2002).

Table 2.3 Low-Carbon Development in the Mexican Electric Power Sector: Barriers and Corrective Actions

Barrier	Corrective action
Large-scale projects	
Planning seeks least-cost technology and does not consider portfolio approach	Modify planning procedures to assess and consider, in addition to costs, volatility risks associated with different technologies, and minimize the portfolio's overall risk and cost over the long term
Planning does not consider ex-plant infrastructure costs and co-benefits	Include other benefits, such as local environmental externalities, all infrastructure costs (for example, ports, pipelines), and possible carbon mitigation revenues
Only large-scale projects can participate in bidding processes	Allow small-scale renewable energy and cogeneration projects to offer partial capacity in bidding processes
Unresolved environmental and social issues associated with large hydro projects	Establish better negotiation mechanisms for planning, construction, and operation of hydropower plants[a]
Small-scale projects[b]	
No predefined contracting procedures for renewable energy and cogeneration projects to sell electricity to the grid	Develop small power purchase agreements
Renewable energy generators only paid short-term marginal costs and not for capacity	Develop payment systems that reward all benefits, including capacity, risk reduction, and externalities (including applicable carbon payments)
No capacity payments for cogeneration projects	Develop payment systems that reward all benefits, including capacity, risk reduction, and externalities (including applicable carbon payments)
Obtaining local and federal licenses is difficult	Establish streamlined licensing procedures
Transmission bottlenecks exist	Expand transmission capacity in areas with large renewable energy potential

Source: Authors.
Note: This table does not consider the new secondary regulations on renewable energy, included in the Anteproyecto de Reglamento de la Ley para el Aprovechamiento de Energías Renovables y el Financiamiento de la Transición Energética.
a. Refer, for example, to the mechanisms proposed by the World Commission on Dams (WCD 2000).
b. Barriers to small-scale projects refer primarily to changes that supply electricity to the grid rather than for self-supply.

Conclusions

Demand for electric power in Mexico has been growing faster than GDP over the past several decades, and this trend is likely to continue, as electricity use continues to grow in all sectors. Meeting the increasing demand for power under a baseline scenario is projected to increase total CO_2e emissions from power generation by 230 percent between 2008 and 2030, from 142 to 322 Mt CO_2e. Based on their economic costs of production—excluding carbon and local externalities—both coal- and gas-fired power generation would increase under the baseline scenario, with coal accounting for 37 percent and gas 25 percent of the new capacity.

Cogeneration could provide about 12.5 percent of new capacity under a low-carbon scenario, at costs that are substantially lower than the current marginal costs of power generation in Mexico. The generation of electricity from biomass is a promising technology for Mexico, with estimated costs that are also lower than current marginal costs. At a cost of CO_2e of up to \$10/t, additional low-carbon energy technologies—hydro, wind, geothermal, and other biomass, such as biogas and bagasse—could replace much of the incremental fossil fuel generation in the baseline scenario. Total incremental investment costs for the MEDEC low-carbon scenario for the power sector amount to \$10 billion between 2009 and 2030, much of which would be offset by lower operation costs.

Despite regulatory mechanisms that favor the development of self-supply renewable energy projects, the environment to tap cogeneration and renewable energy remains inadequate in Mexico. Several policy and regulatory changes are needed to overcome barriers that have inhibited the successful development of the country's renewable energy resources and cogeneration potential. These include low planning prices (including the lack of externalities) CFE assumes for new fossil fuel–based power generation, the lack of recognition of the portfolio effect in planning, and the inability to adjust procurement procedures to the particularities of renewable energy projects.[7] For cogeneration—which has linkages to both the oil and gas sector and other industries in end-use energy—new contracting procedures are needed for small power producers to reduce the risks and transaction costs.

In November 2008, Mexico passed new legislation to promote renewable energy (LAERFTE 2008) as part of the energy reform package, and the corresponding secondary regulations were published in September 2009. Its impact will depend on the methodologies and regulatory instruments that are issued by the Regulatory Commission and SENER in the coming months.

Notes

1. As of October 11, 2009, LyFC has been taken over by CFE.

2. This figure corresponds to projections made in 2007 that are included in the Electricity Sector Outlook 2007–2016. Given the global financial crisis, the rate of growth of the economy may be below this average in the coming years.

3. About 40 percent of this investment will be needed for generation. See SENER (2007) and CFE (2008b).

4. The government's Electricity Sector Outlook 2008–17 (SENER 2008c) sets lower capacity targets, in light of the international financial crisis and the current overcapacity of Mexico's power generation system. The latest projections have electricity demand growing at 3.3 percent per year from 2008 to 2017, and the projected annual GDP growth rate is 2.3 percent. The baseline scenario could be revised to match these recent developments, although given that the overall magnitude of the interventions in terms of tons of CO_2e would be similar, doing so is not necessary.

5. Mexico's electricity law mandates least-cost procurement of electricity genera-
    tion sources. This mandate, as well as its rather strict interpretation by CFE, has
    constituted a barrier to the penetration of cleaner technologies.

6. Although there will undoubtedly be technological change in power-generation
    technologies during the coming two decades, the study takes a technology-
    neutral stance and allows cost reductions from economies of scale only.

7. The analysis of these five interventions was carried out by the electricity team,
    with the collaboration of the energy-efficiency team. A detailed description of
    the assumptions used in the analysis of these interventions is included in appen-
    dix C.

8. In order to foster energy source diversity in the power sector, the Energy Minis-
    try (SENER) has established a 40 percent ceiling for natural gas capacity and a
    25 percent floor for renewable energy capacity, including large hydro. However,
    given the increasing volatility in oil and natural gas prices and the country's
    high dependence on these hydrocarbons, a more effective approach might be
    a planning methodology that considers fuel price volatility, such as the use of
    portfolio theory.

CHAPTER 3

Oil and Gas

The potential to reduce greenhouse gas emissions in Mexico's oil and gas sector through both low-cost and no-regrets interventions is significant. Specific interventions that have good economic rates of return include reducing gas distribution leakage, exploiting cogeneration potential at Petróleos Mexicanos (Pemex) facilities, and improving the efficiency of energy use at refining and processing facilities.[1] The success of Pemex's plans to reverse the decline in oil production and further increase gas production will also play a major role in future greenhouse gas emissions from Mexico, because the alternative is the likely increase in imports of fossil fuels, including coal.

The oil and gas industry in Mexico is a major source of revenue, employment, and national pride. Since being nationalized, in the late 1930s, the oil industry has contributed enormously to the country's development.

Pemex is currently among the largest companies in the world in terms of assets. It is the largest source of export earnings for Mexico and directly employs more than 130,000 people. Although Pemex's contribution to the economy has declined in the past two decades—it accounted for 6.5 percent of GDP in 2008—oil revenues still account for more than one-third of the federal budget.

Among the greatest challenges facing Mexico's oil industry is the need to reduce the decline in oil production. Crude oil production increased from 3.0 million barrels a day (mbd) in 2000 to a peak of 3.4 mbd in 2004. By August 2009, however, production had fallen to about 2.6 mbd, led by the rapid decline in production from Mexico's largest field, Cantarell. As recently as 2004, Cantarell accounted for nearly two-thirds of Mexico's total oil production (2 mbd); since then, production has declined sharply. In July 2009, production at Cantarell was only about 600,000 bd. Production is likely to fall by 15 percent a year between 2009 and 2012. If production from new fields cannot offset the losses from Cantarell, Mexican oil production could fall below 2.5 mbd by 2010, with a resulting large drop in oil exports and a consequent fall in public revenues. Mexico also recognizes the need to improve the efficiency of Pemex.[2]

Recent reforms in the oil and gas industry are intended to provide additional budgetary and financial flexibility to Pemex. Spending by Pemex, a decentralized federal agency, falls under the restrictions of the federal budget, and its financial obligations fall under public borrowing structures.[3] Over the past two decades, a limited federal budget and constraints on borrowing have led to insufficient investment in the oil and gas sector in order to meet production targets and related product quality improvements. Pemex is currently the world's most indebted oil company (total debt was $46.1 billion debt in 2007, and the ratio of debt to proven reserves was $3.1 dollars per barrel of oil equivalent) (figure 3.1). This high level of indebtedness has limited the company's ability to raise financing in private capital markets. The relationship between investment in the oil sector and future energy production and earnings is recognized in Mexico. The problem is the fact that investments in the energy sector compete with pressing social programs, such as health, education, and poverty alleviation, which have relied on oil earnings to finance increases in budget allocations.

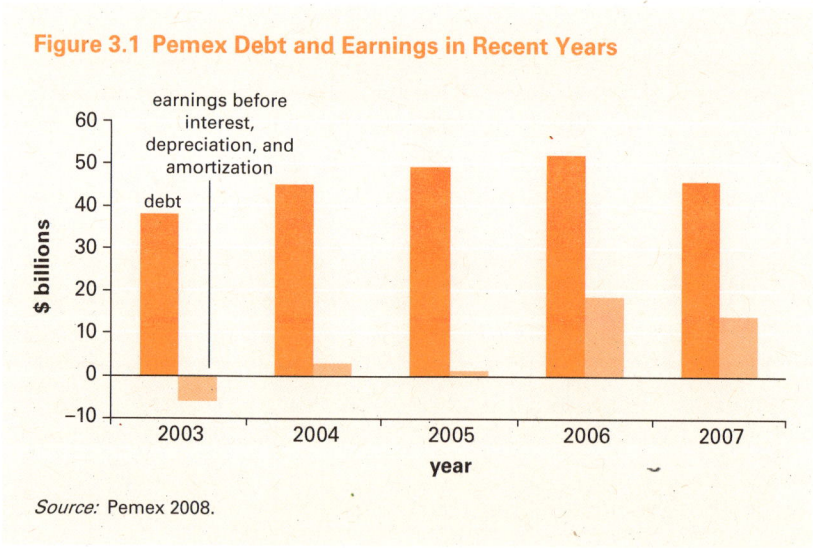

Figure 3.1 Pemex Debt and Earnings in Recent Years

Source: Pemex 2008.

Sustaining and expanding natural gas production is critical to meeting Mexico's energy demand. Gas demand in Mexico has been increasing over the past two decades, as the country expands the use of efficient and clean combined-cycle gas for power generation. Between 2000 and 2007, production of natural gas increased from 4,679 million cubic feet per day (mcfd) to 6,058 mcfd (figure 3.2). The majority of the increase in production has been attributable to the increase in nonassociated gas (gas produced independently of oil). However, the 29 percent increase in production between

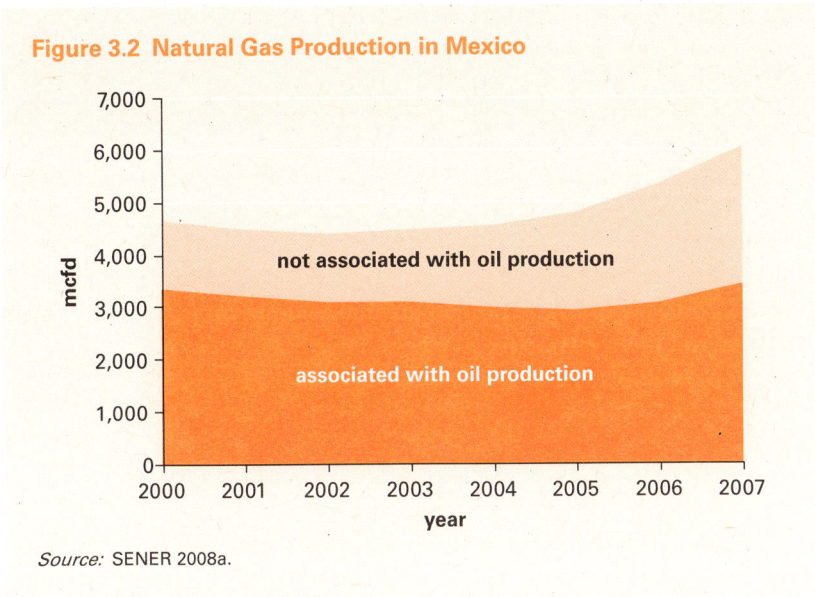

Figure 3.2 Natural Gas Production in Mexico

Source: SENER 2008a.

2000 and 2007 was insufficient to satisfy the increase in demand, which rose 38 percent over the same period. This led to a significant increase in imports of gas, mainly from the United States, a trend that is likely to continue in the near to medium term. At the same time, significant quantities of natural gas are being vented and flared at oil production facilities, principally in offshore areas. If tapped for consumption (as opposed to reinjection, which is another option) and the challenge of high nitrogen content could be overcome, this amount of natural gas could nearly offset natural gas imports. In light of the natural gas situation in Mexico, the Ministry of Energy has set itself the near-term goals of increasing domestic natural gas production and reducing gas flaring and venting.

The Baseline Scenario

Under the baseline scenario, oil and gas production peaks about 2016 and declines thereafter. Energy demand—including gas for power and industry and petroleum products (gasoline and diesel) for the transport sector—is expected to increase throughout this period—exactly when oil and gas production peaks will have a significant impact on the Mexican economy and on greenhouse gas emissions. In the absence of additional domestic gas production, Mexico will need to consume other fuels for power generation. Imported coal is the most likely fuel source based on financial costs and availability. Mexico could also import additional natural gas from the United States or through liquefied natural gas (LNG) projects. Under the baseline scenario, Mexico could cease to be a net energy exporter within the next decade.

The MEDEC Low-Carbon Scenario

Three interventions were evaluated in the oil and gas sector.[4] They include increasing cogeneration in Pemex, improving refinery efficiency, and reducing gas leakage.[5]

Cogeneration in Pemex

Cogeneration potential in Pemex refineries and basic petrochemical plants is equivalent to more than 6 percent of Mexico's total installed capacity. About 3,700 MW of cogeneration potential could be tapped at Pemex's six refineries and four petrochemical plants (table 3.1).[6]

Table 3.1 Pemex Cogeneration Potential

Type of facility	Location	Size of plant (MW)
Refinery	Cadereyta	375
	Madero	350
	Tula	480
	Salamanca	440
	Minatitlán	475
	Salina Cruz	565
Petrochemical plant	Cangrejera	400
	Morelos	300
	Pajaritos	105
	Independencia	200

Source: Pemex 2004.

The operation of refineries and basic petrochemical plants requires considerable volumes of steam, which is generated by burning fossil fuels, such as refinery gas, fuel oil, and intermediate distillates. Modern oil refineries use cogeneration plants to provide steam for refining processes and electricity for both self-consumption and sale to the grid. Cogeneration has become increasingly attractive, with refineries making use of low-value and polluting heavy residual fuel from the refining process, which they clean through gasification. The use of the resulting gasified fuel in cogeneration turbines can result in overall efficiency (thermal efficiency plus electric efficiency), reaching values exceeding 80 percent.[7]

The first stage of investment can be in a combined-cycle cogeneration power station fueled by natural gas. A second stage—which also helps dispose of dirty residual fuel—is to install a gasifier to exploit the residual fuels of the refineries.

Refinery Efficiency

Oil refining is a very energy-intensive industry. Fuel use varies depending on the type of crude oil processed, the mix of outputs produced, and the environmental standards the refined products must meet. In refineries, most conversion processes take place under conditions of high temperatures and pressure, which contributes to the formation of deposits in tubing and equipment that hinders heat transfer, leading to higher fuel consumption. Several methods can be used to reduce the resulting energy losses, including process controls, temperature control, and the cleaning and maintenance of equipment. In some processes it is also possible to recover pressure energy by replacing throttling devices by hydraulic turbines, which in turn drive other machinery or generate electricity (pumps as turbines are especially suitable for this purpose). Reviewing processes, installing new heat recovery systems, implementing maintenance and upgrade practices, and conducting energy development studies and audits can all contribute to improving the energy efficiency of a refinery and reduce greenhouse gas emissions.

In general, the options available to increase energy efficiency in a refinery can be grouped under two large categories: (a) low-cost actions related to energy-management systems that can be implemented in the short and medium term and (b) larger, more comprehensive technological reconfiguration programs, such as investments in fuel use and process technologies that require longer implementation times. Energy-management measures include maintenance programs, installation of heat and pressure recovery equipment, and efficient lighting. Technological reconfiguration programs imply the revision and modification of processes in the refinery, as well as the implementation of more efficient technologies for energy generation, such as energy integration.

As a result of modest energy efficiency measures undertaken by Pemex between 2001 and 2006, energy intensity was reduced 3 percent. Nevertheless, the overall energy efficiency of Mexican refineries remains considerably below international refining industry standards.[8]

To assess the energy-efficiency potential of Pemex refineries, the team evaluated a broad renovation of processes and equipment, including the recovery of hydrogen from exit gases in various process units (hydrocracking, hydrotreating, coking, and fluidized catalytic cracking [FCC]). Given the complexity and size of refineries, it is generally difficult to achieve optimum efficiency through renovation investments. Many investments in Mexican refineries that improve energy efficiency and that are required to meet increasing fuel-quality standards are often not profitable, because it is difficult to pass on the costs of quality improvements to consumers. For this reason, the refinery-efficiency intervention was found to impose net incremental costs relative to the baseline and thus had a positive incremental cost for reducing greenhouse gas emissions. A wide range of less comprehensive, extremely cost-effective investments in existing refineries—for lighting, pumps, motors—can improve energy efficiency.

Gas Leakage Reduction

Reducing losses of natural gas can generate large financial savings. Moreover, because natural gas (methane) has a global warming potential 21 times higher than CO_2, the benefits of reducing methane leakage in terms of carbon payments are among the highest of greenhouse gas mitigation interventions. Methane emissions from natural gas systems account for an estimated 18 percent of total worldwide methane emissions, with Mexico emitting about 7 percent of the global total from natural gas systems.

Nearly 80 percent of methane emissions from natural gas transport in Mexico are associated with wet seals used with the operation of compressors within the production, storage, and distribution network. The replacement of wet seals with dry seals allows the use of high-pressure systems, which can reduce the leakage of methane, require less maintenance, and reduce the risk of accidents. The potential for this technology in Mexico is large, given that 46 of 67 compressors still use wet seals (table 3.2). An economic analysis of replacing wet seals with dry seals was conducted using the Ciudad Pemex gas-processing center as the reference case. Based on the results, it was assumed that the program could be applied in all gas centers with wet-seal compression systems.

Table 3.2 Potential for Compressor Seal Replacement in Mexico's Gas Processing Centers

Center	Number of compressors with wet seal	Number of compressors with dry seal
Cactus	15	0
Nuevo Pemex	11	0
Ciudad Pemex	3	3
Coatzacoalcos	3	0
Poza Rica	4	0
Reynosa	2	0
Burgos	0	18
La Venta	5	0
Matapionche	3	0
Total	46	21

Source: Authors.

Wet-seal replacement was estimated to have a reduction potential of 3 million tons of CO_2e through 2030, or an average of 140,000 tons a year. Recent estimates of natural gas losses in Mexico (as reported by the Methane to Markets program) indicate that losses could be significantly higher than the official figures cited above, in which case measures to identify and implement measures to reduce losses would be of even greater importance for greenhouse gas mitigation policy in Mexico.

Summary of Oil and Gas Interventions

The greatest net benefit comes from increasing cogeneration in Pemex facilities, followed by increasing refining efficiency and reducing gas leakage (table 3.3). Other interventions in the oil and gas sector were considered and assessed but ultimately not included in the MEDEC scenario, because they did not meet the MEDEC criteria, because data were not available, or for other reasons. Reducing gas flaring and venting may be a cost-effective intervention, but Pemex is planning to implement the intervention in the coming years, which means that it has become part of the baseline scenario. Reducing fugitive methane emissions in the oil and gas industry from sources other than gas compression stations, such as oil storage facilities, may be cost-effective, but too few data were available to assess the corresponding potentials and costs. Lack of data prevented a careful cost-benefit analysis of other potential opportunities as well.

Table 3.3 Summary of MEDEC Interventions in the Oil and Gas Sector

Intervention	Maximum annual emissions reduction (Mt CO$_2$e/year)	Net cost or benefit of mitigation ($/t CO$_2$e)
Cogeneration in Pemex	26.7	28.6 (benefit)
Gas leakage reduction	0.8	4.4 (benefit)
Refinery efficiency	2.5	16.6 (cost)

Source: Authors.

Barriers to Mitigating Greenhouse Gas Emissions

Barriers to the implementation of low-carbon interventions in Mexico's oil and gas sector include intervention-specific barriers and barriers that are symptomatic of the organizational and management structure of Pemex (box 3.1). From Pemex's perspective, although investments in cogeneration plants, for example, have excellent rates of return, such investments are less attractive than petroleum exploration and development. They are therefore not a high priority from Pemex's perspective.

Because of Pemex's high debt, it has had difficulty tapping commercial credit markets at reasonable terms. Recent oil industry reforms have been aimed at improving the situation. However, given Mexico's dependence on oil industry revenues for financing the federal budget, reform measures that reduce tax payments by Pemex are likely to be limited in the short term.

The most significant barrier to implementation of cogeneration in Mexico is the unfavorable conditions for the sale of surplus electricity to the grid. Pemex's electricity demand is currently in the range of 900 MW—a fraction of the potential for cogeneration of more than 3,700 MW. Although some of the inefficient electricity production in Pemex can be replaced by more efficient cogeneration, Pemex must be able to sell surplus electricity (as well as the corresponding capacity) to CFE in order to tap the full potential of cogeneration in its facilities. Since the cost of cogeneration from Pemex

Box 3.1 Financing Pemex Infrastructure Projects with High Environmental Benefits

A smaller federal budget and limited borrowing capacity have reduced the ability of Pemex to allocate financial resources to capital projects with a high environmental benefit and a high return in recent years. The higher financial rate of return expected on exploration and development (E&D) activities has precluded the possibility of financing these and other projects, despite their environmental benefit and attractive returns. On average, E&D investment has accounted for more than 80 percent of Pemex's portfolio.

An additional factor hampering the financing of non–E&D projects is Pemex's huge debt (see figure 3.1 and figure below), which reduces the company's ability to raise funds in the commercial finance markets. Pemex has tapped commercial credit in the past for infrastructure investments, but given its poor credit rating, the company has typically used other financing mechanisms (namely, the federally approved budget and PIDIREGAS). Although the ratio of earnings before interest, depreciation, and amortization to debt has been positive in recent years, the international financial crisis may limit the ability of the company to obtain commercial finance for its investments.

Ratio of Total Debt to Proven Reserves for Selected Oil Companies, 2007

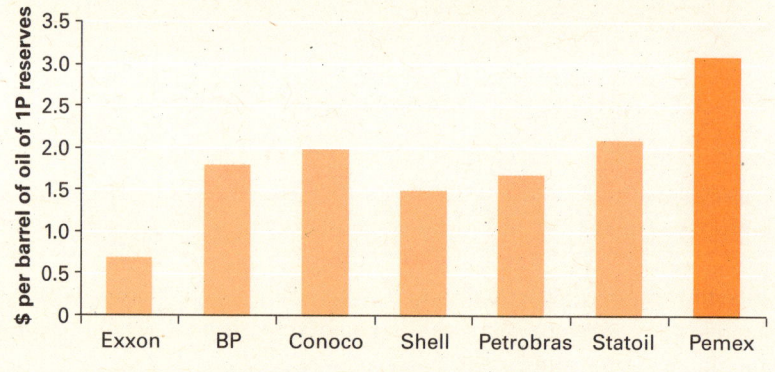

Source: Authors' calculations based on Pemex 2008.

facilities is estimated to be significantly lower than the new power capacity CFE is planning to contract, the benefits to Mexico are clear.

In theory, investment in cogeneration facilities by Pemex could be contracted to the private sector, because it does not involve the "ownership" of oil or gas resources. This is not the case for reducing gas leakage or improving refinery efficiency. Contractual arrangements with private investors would face more onerous legal obstacles in these areas.

There are many reasons why Pemex has not adopted more energy-efficiency measures in its refineries. Many relate to investment restrictions imposed on Pemex by the federal government and the lack of success in upgrading refineries to meet tighter fuel quality standards and dispose of highly polluting residual fuels.

Conclusions

There is significant potential to reduce greenhouse gas emissions in Mexico's oil and gas sector through both no-regrets and low-cost interventions. In particular, there is significant cogeneration potential in Pemex facilities, where more than 6 percent of Mexico's total installed electric power capacity could be installed. Another intervention that can reduce greenhouse gas emissions and that has good economic rates of return is reducing gas distribution leakage.

A number of policy constraints have limited energy-efficiency investments in Pemex. The hope is that passage of the oil industry reform measures in 2008 will make it easier for Pemex to undertake needed investments, including in efficiency improvements. A major limitation remains the extremely large share of the federal budget provided by oil revenues from Pemex. Measures to allow contracting with the private sector to finance cogeneration investments could overcome some of the current investment constraints for Pemex, within existing Mexican laws.

Overcoming barriers to domestic gas production will be a key determinant of Mexico's future CO_2 emissions, because under all scenarios natural gas will need to increase substantially to meet the growing demand from the electric power, industrial, and residential and commercial sectors. Without large new sources of natural gas, the least-cost alternative for power generation will be coal.

One of the objectives of the energy reform program approved by the Mexican congress in 2008 was to provide greater flexibility to allow Pemex to operate in a manner similar to other national oil companies. Despite greater financial, budgetary, and procurement flexibility included in the reforms, it is still unclear whether the private sector will be attracted to Pemex's new terms, especially in E&D. Although the reform does not allow private investment in downstream activities, the hope is that Pemex will be able to contract services provided by the private sector. Considering that some low-carbon investment projects, such as cogeneration, could be provided by service contracts with Pemex, these projects could help test the effect of the recent reforms for improving investment in non–E&D activities.

Notes

1. For the purposes of this study, the oil and gas sector includes the extraction of oil and gas; the refining, transport, and distribution of oil products; the transport, processing, and distribution of natural gas; and a portion of secondary petrochemical production in Mexico coming from Pemex facilities.

2. A number of countries have moved to improve the investment climate and the accountability of their state-owned energy companies over the past two decades.

Petrobras of Brazil and Statoil of Norway, for example, have modernized their oil industries, making them among the most efficient and profitable in the world.

3. Recent reform legislation also ended the use of PIDIREGAS for long-term financing for Pemex (see chapter 2).

4. The analysis of all interventions except cogeneration was carried out by the energy efficiency and oil and gas team. The electricity team analyzed cogeneration.

5. The reduction in gas flaring and venting is an important mitigation strategy in the oil and gas sector. As of 2007, Mexico was among the largest gas-flaring countries in the world, with a total of 5.6 billion cubic meters and an emission rate that is high by international standards. Most of Mexico's unrecovered associated gas (gas that is produced as a by-product of oil production) is flared offshore at the Cantarell field. Gas flaring and venting is estimated to have produced up to 44 Mt CO_2e in 2007, which accounts for as much as half of total greenhouse gas emissions from the oil and gas sector (extraction, refining, and production of oil and gas), or about 6 percent of total national emissions. Pemex is currently undertaking investments to substantially reduce flaring and venting by 2012; because of this strategy, it is not included as a MEDEC intervention.

6. Pemex recently launched the bidding process for the construction and operation of a cogeneration plant at the Nuevo Pemex gas-processing plant, and plans are nearly complete for a similar investment at the Salamanca oil refinery. In addition to producing heat to satisfy the gas plant, the facility will provide 300 MW of power capacity to cover the electricity requirements of this and other Pemex facilities in southeast Mexico (by wheeling power through the electricity grid).

7. The efficiencies for the whole process (from residual fuels to heat and power) are somewhat lower.

8. The Solomon Index of Energy Efficiency, an international parameter of refinery efficiency, was improved for Mexican refineries from 122 in 2001 to 118 in 2006. For comparison, the average efficiency for Canadian refineries is 93 on this scale.

CHAPTER 4

Energy End-Use

Managing the growth of electricity and fuel demand through energy-efficiency measures in the end-use sectors will be critical to mitigating Mexico's greenhouse gas emissions. The industrial, residential, and commercial and public service sectors account for 95 percent of electricity consumption in Mexico, and their electricity use has been growing at more than 4 percent a year since 1995. By contrast, the fuel consumption (direct consumption of gas, oil products, and coal) of these three sectors, which account for about 42 percent of total end-use fuel in Mexico, has remained essentially flat since 1995. These trends reflect the changing production pattern in the industrial sector (away from fuel-intensive basic materials), as well as the impact of rising wealth in urban areas, which tends to drive up electricity consumption.

This chapter examines the contribution of stationary (nontransport) energy end-use to Mexico's greenhouse gas emissions and the potential and cost of emissions reduction through energy-efficiency improvements. The analysis focuses on the three large end-use sectors (as defined by national energy statistics): industry, residential, and commercial and public services. Together these sectors account for about 48 percent of total energy end-use in Mexico (figure 4.1). (Transport, which depends almost exclusively on oil products, accounts for 49 percent of energy end-use, and is analyzed in chapter 5.)

Mexico's national energy-efficiency programs started in the early 1990s, after the establishment of Comisión Nacional para el Ahorro de Energía (CONAE) (National Commission for Energy Savings) in 1989 and FIDE in 1990. Following passage of the Law for Sustainable Energy Use in November 2008, the Comisión Nacional para el Uso Eficiente de la Energía (CONUEE) (National Commission for the Efficient Use of Energy) was established as an administrative body with technical and operational auton-

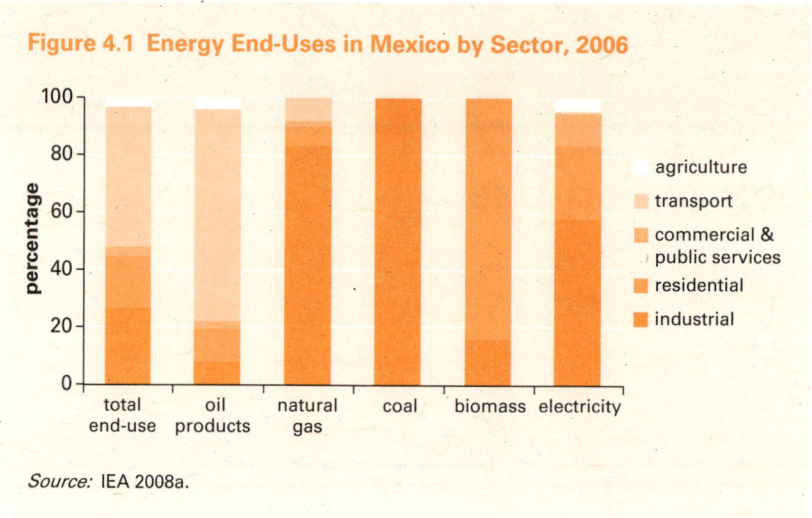

Figure 4.1 Energy End-Uses in Mexico by Sector, 2006

Source: IEA 2008a.

omy. CONUEE's mission is to promote energy-efficiency and serve as the technical authority for the sustainable use of energy. FIDE, a private-public trust fund created by the federal electricity utility CFE, has been a leader in promoting electricity savings through demand-side management measures, such as the introduction of compact fluorescent lamps and the retirement of old appliances. It is estimated that as of 2006, standards related to electricity end-uses saved an accumulated total of 16,065 GWh and avoided about 2,926 MW of generation capacity. FIDE energy-efficiency programs achieved estimated electricity savings of 15,146 GWh and avoided 1,745 MW of generation capacity as of 2008 (FIDE 2008).

There remains considerable potential for energy-efficiency improvements in Mexico. After significant improvements in the 1990s, the downward trend in Mexico's energy intensity of GDP has stalled (figure 4.2), primarily because of the rapid increase in electricity consumption, which has grown significantly faster than GDP. Both CONUEE and FIDE have set ambitious targets for electricity savings by 2012. The baseline analysis provides the context for understanding the energy savings potential in each of the three major end-use sectors.

The Baseline Scenario

The industrial, residential, and commercial and public sectors account for the majority of electricity use and a substantial share of other fuel use in Mexico. The industrial sector is characterized by both extremely modern and energy-efficient industries, such as steel and cement, and antiquated and high energy-consuming industries, many of them small and medium-size firms. In the residential, commercial, and public sectors, the demand for air conditioning and refrigeration has been increasing and is likely to continue to do so as incomes rise. Room for growth is significant, as per capita

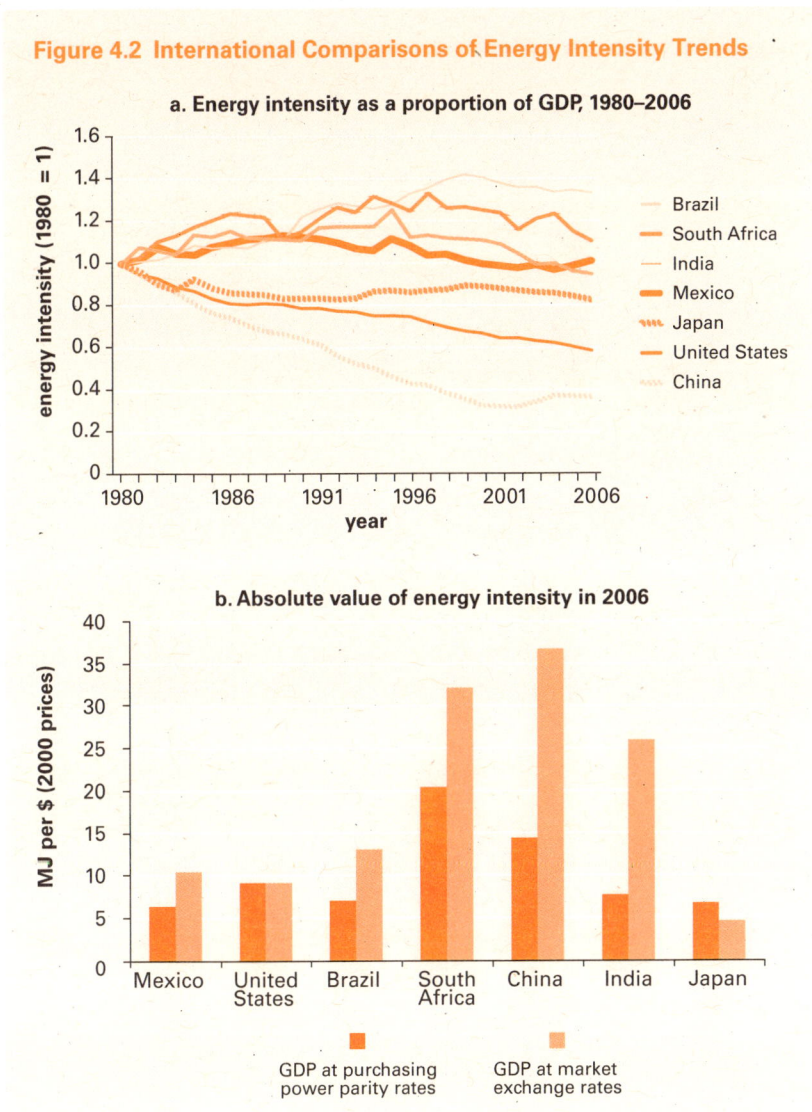

Figure 4.2 International Comparisons of Energy Intensity Trends

a. Energy intensity as a proportion of GDP, 1980–2006

- Brazil
- South Africa
- India
- Mexico
- Japan
- United States
- China

b. Absolute value of energy intensity in 2006

- GDP at purchasing power parity rates
- GDP at market exchange rates

Source: Based on data from the U.S. Energy Information Administration (www.eia.doe.gov).

electricity use in Mexico remains a fraction of that in high-income countries with similar climates.

The Industrial Sector

The industrial sector is the second-largest energy end-user in Mexico (after transport), accounting for about 27 percent of total energy end-use.[1] It is the largest electricity user, accounting for 58 percent of total electricity consumption.[2] More than half the industrial energy use is in five main subsectors, which also account for the majority of fuel use (oil products, gas, solid fuels): cement (nonmetallic minerals), iron and steel, chemicals and petrochemicals, mining, and food and tobacco (figure 4.3).

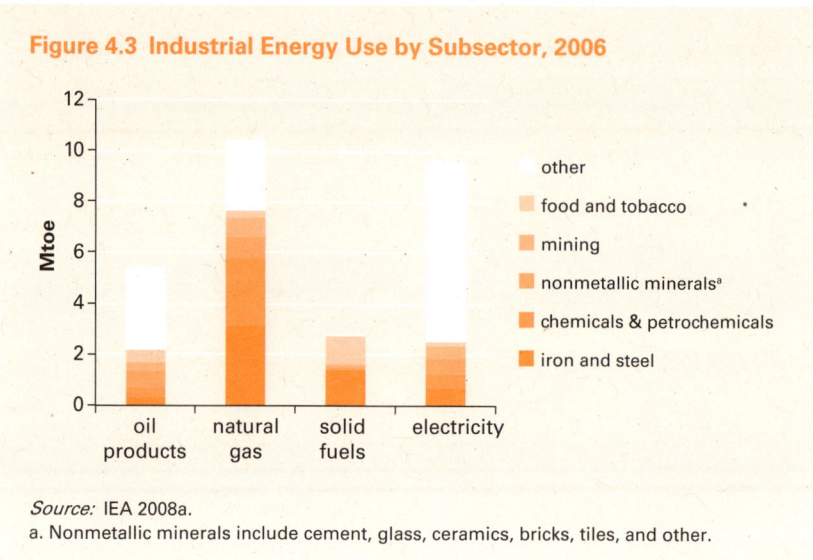

Figure 4.3 Industrial Energy Use by Subsector, 2006

Source: IEA 2008a.
a. Nonmetallic minerals include cement, glass, ceramics, bricks, tiles, and other.

Some large-scale basic materials industries in Mexico are relatively efficient by international standards. Mexico's iron and steel industry, for example, is among the least carbon intensive in the world, thanks in part to its reliance on advanced technologies.[3] The energy intensity of crude steel in Mexico has remained below 14 gigajoules per ton (GJ/t) since the early 2000s, compared with the global average of about 20 GJ/t. In Mexico's cement industry, total primary energy use per ton is 19 percent lower than in Canada and 27 percent lower than in the United States, albeit about 15 percent higher than that of world leaders Brazil and Japan (IEA 2007). Nonetheless, a large portion of Mexico's industrial sector is made up of small and medium enterprises in a wide range of activities that have relatively high energy intensity. They often use older equipment and lack access to technical know-how and financing for upgrades.

Broadly speaking, in addition to cogeneration, the main sources of energy savings in the industrial sector come from energy-efficiency improvements in motor systems, steam systems, and kilns and furnaces. Motor systems account for 70 percent of total industrial electricity consumption in Mexico, and steam systems are responsible for an estimated 40 percent of industrial fuel consumption. Kilns and furnaces account for most of the remaining industrial fuel and electricity consumption. According to the International Energy Agency's industrial energy-efficiency assessment (IEA 2007), adopting international best practices would reap technical energy savings of 20 percent in industrial motors, 10 percent in steam systems, and 15 percent in kilns and furnaces. About 80 percent of Mexico's industrial cogeneration potential has not been utilized; the undeveloped potential is concentrated in Pemex facilities (see chapter 3) and in the food processing, chemical and pharmaceutical, automobile, pulp and paper, textile, glass, and sugar industries.

The Residential Sector

The residential sector accounts for about 18 percent of total energy end-use in Mexico. Its share of total electricity consumption increased from 16 percent in 1995 to 22 percent in 2006. Per capita residential electricity consumption in Mexico is about 320 kWh/year—about one-tenth of the 3,150 kWh/year consumed in the United States. In the states of Arizona, New Mexico, and Texas, which have high air-conditioning demand and a climate that is similar to that of large parts of Mexico, electricity accounts for up to 80 percent of residential energy consumption. As incomes grow in Mexico, the implied growth potential for residential electricity demand is staggering. In urban areas of Mexico, cooking and water heating rely primarily on liquefied petroleum gas (LPG), which accounts for more than 53 percent of residential fuel consumption.

Biomass consumption. Biomass consumption, which accounts for about 40 percent of residential fuel, has remained stable in Mexico; it is used primarily by rural households for cooking in traditional open fires. The residential use of biomass is relevant for greenhouse gas emissions for two primary reasons. First, biomass consumption produces net CO_2 emissions, because a portion of the fuelwood used is not harvested in a sustainable manner. Second, non-CO_2 gases are emitted because of incomplete biomass combustion. In addition, the traditional use of biomass is linked to severe respiratory and other health problems, especially among women and children in rural households, because of exposure to smoke from inefficient fuelwood combustion. The experience in Mexico shows that the transition to LPG among rural households faces important economic and cultural barriers; in the short term, improving biomass stoves is a more feasible way to address both health impacts and greenhouse gas emissions (Troncoso and others 2007).

Air conditioning, refrigeration, and home appliances and electronics. Air conditioning, refrigeration, and home appliances and electronics are expected to be the main growth areas of residential electricity demand in Mexico. Currently, these three end-uses plus lighting account for about-equal shares of residential electricity consumption. Air-conditioner saturation rates in Mexico were only about 20 percent in 2005, compared with about 95 percent in regions of the United States with similar cooling-degree days. One recent study projects that air-conditioner electricity use in Mexico could increase 10-fold by 2030, reaching a value that is three times higher than total residential electricity use in 2005 (McNeil and Letschert 2008). The saturation rate of refrigerators is relatively high in Mexico, at 82 percent (2006), but it still has room to grow, both in number and storage capacity. Recent efforts to promote compact fluorescent lamps notwithstanding, incandescent lamps still account for about 85 percent of the in-use residential light bulbs in Mexico, indicating a large potential for scaling up use of compact fluorescent lamps.

Mexico has minimum energy performance standards (MEPS) for 18 types of electricity-consuming equipment, including air conditioners, refrigerators, and clothes washers. In general, these standards are on par and consistent with the MEPS in the United States, because of harmonization efforts begun in the early 1990s. Large electricity savings can be achieved through the accelerated retirement of old and inefficient air conditioners and refrigerators and the enforcement of increasingly stringent mandatory MEPS on new products. The availability of cheap and inefficient secondhand air conditioners from the United States is a particular problem for northern Mexico, where air-conditioning demand is also highest.

Mexico does not have a residential building energy-efficiency code. Such a code has proven to be a highly effective means of reducing cooling loads (through thermal insulation and window improvements) in the U.S. state of California, which has progressively pursued mandatory building energy-efficiency codes since the late 1970s. The combination of codes for residential buildings with inherently lower cooling demands and high-efficiency air conditioners can drastically reduce air-conditioning electricity consumption in new homes.

Domestic hot water accounts for about 52 percent of residential LPG and natural gas consumption in Mexico; it is the main end-use driving up residential fuel consumption (PROCALSOL 2007). Although there is potential to improve the energy efficiency of hot water boilers, much larger fossil fuel savings can be achieved by scaling up the application of solar water heaters, especially in low-density dwellings, such as single-family homes and townhouses.

The Commercial and Public Services Sector

Energy use by the commercial and public services sector in Mexico is estimated to account for less than 4 percent of total energy end-use. The sector is nevertheless an important electricity consumer, accounting for more than 21 percent of total electricity use.[4] As cities expand and modernize, the commercial and public services sector will assume a much larger role in Mexico's energy use. In the United States, the commercial and public sectors account for about 14 percent of total energy end-use and 35 percent of total electricity consumption.

Lighting accounts for more than half of electricity consumption in the commercial and public sector in Mexico; air conditioning and refrigeration account for about 18 percent each; and the energy used by water supply and sanitation companies accounts for about 9 percent. As a large portion of the commercial and public services sector (public buildings and municipal water companies) is owned by federal, state, or municipal governments, substantial economies of scale are available through fairly simple procurement and retrofit programs.

New commercial buildings are subject to two national standards enforced through third-party verifications. The lighting system standard is enforced through the service contracting process of the national utilities (CFE and

LyFC), which require compliance certificates for the provision of service. The compliance for the building thermal envelope standard has to be mandated by local codes and enforced by local authorities. A lack of local capacity and political will has contributed to significant lapses in enforcement of the building envelope standard. Given the large growth potential of electricity use in the commercial and public services sector, a focus on tightening the energy-efficiency standards and enforcement for lighting, refrigeration, air conditioning, and buildings will be crucial to reducing greenhouse gas emissions from this sector.

Energy End-Use Demand Projections

In the baseline scenario, electricity demand is projected to reach 425 TWh by 2030, up from 222 TWh in 2008 (excluding transmission and distribution losses, nontechnical losses, and in-plant consumption). The combined contribution of the residential, commercial, and public service sectors is projected to increase to 67 percent, up from 47 percent in 2008.

The most important component of end-use fuel consumption is transport, discussed in chapter 5. For the industrial, residential, commercial, and public sectors, fuel demand is estimated to increase at an average annual rate of less than 2 percent (figure 4.4).

Figure 4.4 Energy End-Use by Sector: Baseline Scenario

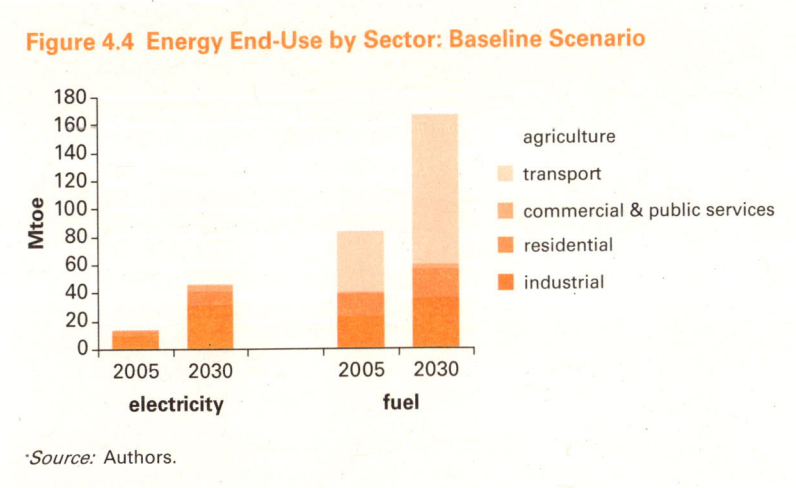

Source: Authors.

The MEDEC Low-Carbon Scenario

The MEDEC study evaluated the costs and impact of 11 energy end-use interventions. Each is briefly described in the following paragraphs.[5]

Electricity End-Use Efficiency

Residential Air Conditioning. This intervention focuses on the 1 million households in Mexico with the highest air-conditioning use. It entails accelerating the phaseout of old air conditioners by 2030 and installing thermal insulation in these households. The combined effect of the new standard-compliant air conditioners and the thermal insulation is assumed to reduce air-conditioning electricity consumption by these households from 4,000 kWh/year to 700 kWh/year.

Residential Lighting. In 2008 there were an estimated 234 million light bulbs operating in about 29 million households in Mexico. The MEDEC scenario assumes that 85 percent of all light bulbs used one hour a day or more in 80 percent of households will be compact fluorescent lamps.

Residential Refrigeration. This intervention proposes the accelerated substitution of refrigerators 10 years or older by new devices compliant with current standards.

Nonresidential air conditioning. The air-conditioning electricity consumption of commercial and public sector buildings was estimated for several types of commercial and public buildings. This intervention assesses the effect of accelerating the substitution of air conditioners in these buildings with advanced devices.

Nonresidential lighting. This intervention involves accelerating the substitution of low-efficiency fluorescent lighting with high-efficiency T8 lighting.[6]

Street lighting. This intervention proposes substituting the entire stock of mercury vapor, incandescent, halogen (iodine-quartz), and fluorescent street lamps by high-efficiency high-pressure sodium lamps.

Industrial motors. This intervention involves the accelerated substitution of large industrial motors and the introduction of high-efficiency (above the current standard) motors. Although the efficient motors are more than twice as expensive as standard motors, the intervention produces net economic benefits.

Cogeneration

Cogeneration in industry. The estimated potential for cogeneration in Mexican industry is about 6,800 MW, excluding the oil and sugar industries. This potential is concentrated in industries with steam requirements in which topping-cycle plants can be used. It is a conservative estimate, as it excludes medium- and small-scale cogeneration schemes. The conditions for bottoming-cycle plants are less favorable, because the waste heat from such sectors as cement and steel and iron is of too low a temperature to be utilized efficiently (CONUEE 2009).[7] Cogeneration enables the construction of new power capacity by utilities to be delayed, leading to higher overall efficiencies in the energy system.

Bagasse cogeneration. Low-efficiency cogeneration plants are currently in operation in most sugar mills in Mexico, fueled by a mixture of bagasse and fuel oil. By substituting these plants with high-pressure, high-efficiency plants, sugar mills can deliver surplus electricity to the grid and cease using fuel oil.

Renewable Heat Supply

Solar water heating. This program entails increasing the penetration of solar water heaters to reduce the use of LPG or natural gas in both existing and new homes. It is assumed that by 2030, 80 percent of new households and 60 percent of households existing in 2008 will have installed solar water heaters.

Improved cookstoves. This intervention entails replacing traditional open fires by more efficient devices in rural households. Penetration by 2030 is assumed to reach 100 percent of rural people who use traditional open fires. Improved cookstoves reduce fuelwood consumption and improve combustion efficiency, thereby reducing both the net CO_2 emissions linked to the nonrenewable fraction of biomass and the non–CO_2 emissions linked to incomplete combustion. At least two government programs and several nongovernmental projects are currently operating in Mexico, providing reason for optimism that the assumed penetration rate can be achieved. Doing so would require, however, that appropriate training, technical assistance, and follow-up be provided, as most ongoing programs provide funds only for the purchase and installation of cookstoves. This intervention produces large net benefits when health and time savings benefits are included (box 4.1 and figure 4.5).[8]

Table 4.1 summarizes the energy end-use interventions, almost all of which are no-regrets interventions. Other interventions in the energy end-use sectors were considered and assessed but ultimately not included in the MEDEC scenario, because they did not meet the MEDEC criteria, because data were not available, or for other reasons. In particular, the pumping

---

Box 4.1 Reducing Emissions, Saving Time, and Providing Health Benefits through Improved Cookstoves

Improved cookstoves are a cost-effective tool for reducing greenhouse gas emissions even without valuing the time family members save by not having to collect as much fuelwood and the health benefits from reduced indoor air pollution impacts. When time savings and the positive health impacts of reducing exposure to fine particulate matter (PM2.5) and carbon monoxide are considered, the intervention provides major benefits to households and society. The net benefit of the intervention rises from essentially zero to $2.34/t CO_2e when time savings are included and to $18.90 when both time and health benefits are included. With about 80 percent of the rural population in Mexico dependent on wood for cooking and heating (Armendáriz and others 2008), the greenhouse gas mitigation potential of widespread introduction of improved cookstoves is substantial.

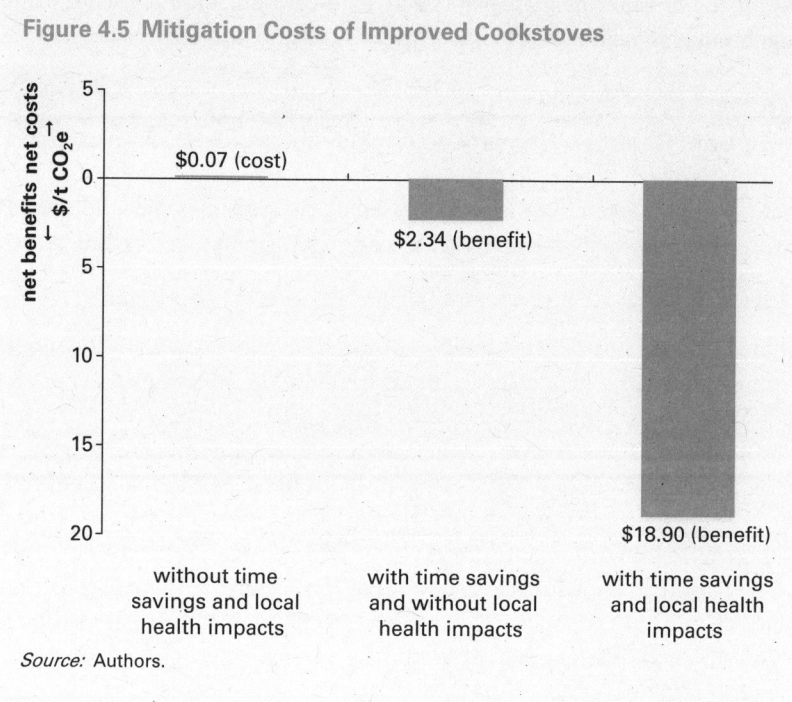

Figure 4.5 Mitigation Costs of Improved Cookstoves

Source: Authors.

Table 4.1 Summary of MEDEC Interventions in the Energy End-Use Sectors

Intervention	Maximum annual emissions reduction (Mt CO$_2$e/year)	Net cost or benefit of mitigation ($/t CO$_2$e)
Electricity end-use efficiency		
Residential air conditioning	2.6	3.7 (benefit)
Residential lighting	5.7	22.6 (benefit)
Residential refrigeration	3.3	6.7 (benefit)
Nonresidential lighting	4.7	19.8 (benefit)
Nonresidential air conditioning	1.7	9.6 (benefit)
Street lighting	0.9	24.2 (benefit)
Industrial motors	6.0	19.5 (benefit)
Cogeneration		
Cogeneration in industry	6.5	15.0 (benefit)
Bagasse cogeneration	6.0	4.9 (cost)
Renewable heat supply		
Solar water heating	18.9	13.8 (benefit)
Improved cookstoves	19.4	2.3 (benefit)

Source: Authors.

of water for irrigation, water supply, or drainage purposes has a significant mitigation potential, and a number of pressure-recovery opportunities could be harnessed by means of hydraulic turbines. Lack of adequate data prevented the thorough examination of these interventions.

Barriers to Mitigating Greenhouse Gas Emissions

The barriers to improving energy end-use efficiency are understood; various barrier-removal policies and instruments have had successes (table 4.2). The approaches and processes to barrier removal are often as varied as the country or locality in which they are applied.

Table 4.2 End-Use Efficiency: Barriers and Corrective Actions

Barrier	Corrective action
Industrial and commercial sectors	
Limited awareness of energy efficiency, including costs, benefits, and risks of new technologies and actions	Industry awareness campaigns on energy-efficiency opportunities, technology seminars and expositions
Few examples presenting the business case for energy efficiency, limited market data, and few identified opportunities to encourage private sector participation	Development and dissemination of targeted energy-efficiency information, technical guides, case studies, project databases, and benchmark studies
Lack of expertise to conduct quality audits and identify energy efficiency opportunities, lack of market expertise to package investments into bankable project proposals	Technical training of energy managers, ESCOs, and auditors; development of standardized template audit reports, bidding documents, and case studies
High import tariffs for energy-efficiency equipment	Establish tax waivers and/or incentives for energy-efficiency equipment purchases
Low or questionable quality of energy-efficiency equipment	Update/expand energy-efficiency standards, labels, and codes
High project development costs (audits) and transaction costs	Develop standard loan procedures, monitoring and verification protocols, and bidding documents; dedicate funds for energy-efficiency audits
Limited private sector investment in energy efficiency (for audits, advisory services, leasing, ESCOs) due to limited equity and available financing	Develop local business models/ESCOs, promote joint venture options and venture capital funds, make small grants to stimulate the market and ESCOs
Limited banking expertise to assess energy-efficiency proposals, low-quality loan applications, high perceived risks for energy efficiency projects	Provide technical assistance to local financial institutions, and conduct demonstrations of project performance
Unclear responsibilities and incentives among building developers, owners, and tenants (principal-agent problem)	Improved building codes/certificates, incentives for green buildings, energy metering
Poor customer creditworthiness or limited debt capacity among borrowers	Create dedicated financing schemes (revolving funds, pooled financing), credit enhancement mechanisms, and alternative financing models to share risks for energy-efficiency projects

(continued)

Table 4.2 End-Use Efficiency: Barriers and Corrective Actions *(continued)*

Barrier	Corrective action
Public sector	
Limited awareness of energy efficiency, including its costs, benefits, risks, and service options	Awareness campaigns targeted to public administrators, case studies
Limited incentives to implement energy-efficiency projects (due to potential loss of budget) and to explore new approaches	Revise budgeting to allow retention of energy savings, awards for agencies/public staff that improve energy efficiency
Restrictive budgeting, financing, and procurement and contracting rules	Revise public policies to encourage energy-efficiency products (for example, life-cycle costing) and ESCOs, develop alternative ESCO models to suit local conditions, create dedicated energy-efficiency revolving funds for public agencies
Residential sector	
Limited awareness of energy efficiency, including its costs, benefits, and risks	Public energy-efficiency awareness campaigns
Concerns over which products are energy efficient and about actual costs/savings, quality, reliability, high upfront investment costs, high transaction costs	Update/expand energy-efficiency standards, labels, and codes; conduct manufacturer negotiations; seek market transformation (through bulk purchase, for example); energy costs/savings information provided through utility billing, utility-financed energy-efficient investments
Low energy pricing	Energy sector pricing and institutional reforms
Unclear responsibilities and incentives among building owners, developers, and tenants (principal-agent problem)	Improved building codes/certificates, incentives for green buildings

Source: Authors.

The Industrial and Commercial Sectors

For the industrial and commercial sectors, the most common issue is institutional. Decision makers, including senior management and financial officers, typically have other investment priorities, such as essential maintenance and repair, production expansion, and product quality enhancements. They therefore give very low priority to investments in reducing operating costs. Many countries have developed energy service company (ESCO) business models, often alongside dedicated energy-efficiency funds, which allow company managers to pay from energy savings, and thus do not need to alter investment priorities or take on additional technical and performance risk. Unfortunately, industrial country ESCO models, which rely on detailed legal contracts, have often been too complex for many developing countries to implement and have not proven viable. There is growing experience with developing local ESCO models, which are having more success (see Taylor and others 2008).

A number of efforts have been initiated in Mexico since 2004 to promote market-oriented energy efficiency, mostly focusing on dedicated energy-efficiency financing programs.[9] Although there have been some innovative and promising proposals, a fundamental gap has been the underlying business model needed to support the transaction. There is often a mistaken belief that if a bank determines that a customer is not creditworthy, the ESCO can finance the project. The reality is that ESCOs are generally unable and unwilling to take on both project performance and credit risks, especially if the customer is deemed risky, despite the attractiveness of the underlying project. Furthermore, new ESCOs in developing countries often have limited experience and weak balance sheets, which may not be capable of handling the full performance risks and dealing with high project development and monitoring and verification costs. Developing more robust, Mexican-grown models, along with financing programs designed based on these models, is much more likely to result in meaningful investments in energy efficiency in the industrial and commercial sectors.

The Residential Sector

Electricity subsidies for middle- and high-income residential consumers discourage many energy-efficiency investments in appliances and lighting. Most electricity consumers in Mexico receive some subsidies; residential and agricultural consumers are the most heavily subsidized (box 4.2).

The residential sector is complex, given the diverse nature of the sector, the large numbers of households, and often limited disposable income. A major issue is the high implicit discount rates households often use when considering energy-efficiency investments. In addition, if the additional cost of the more efficient appliance or equipment is very high, it is less likely to be adopted, regardless of the life-cycle cost. Programs that can reduce costs (for example, bulk purchase, manufacturer negotiations, subsidies, rebates), ensure product quality and cost-effectiveness, and provide an efficient and effective distribution mechanism have a good track record. Mexico already has strong experience with implementing residential appliance programs and has faced these difficulties in the past. Expansion of such programs, particularly targeting air conditioning, lighting, and solar water heating, could have significant impacts.

Several programs and projects in Mexico focus on the dissemination of improved cookstoves (see box 4.1). Potential barriers to large-scale implementation include resistance from rural and indigenous communities because of established traditions and habits, the lack of standardized construction techniques, the difficulty of reaching a large and dispersed rural population, the lack of trained personnel in both the social and technical aspects of cookstove dissemination, and high follow-up costs.

The Public Sector

The public sector faces many procedural and incentive barriers. Government agencies often have restrictive budgets that do not allow them to undertake equipment upgrades; if they do, the financial benefits may not accrue to them. Procurement rules typically favor least-cost equipment

Box 4.2 Underpricing Electricity through Residential Subsidies

The underpricing of electricity to residential consumers in Mexico results in overconsumption and excessive greenhouse gas emissions (Komives and others 2009). Electricity subsidies in Mexico are among the largest in the world ($9 billion in 2006). More than two-thirds of total electricity subsidies go to residential consumers. Average residential electricity prices cover only about 40 percent of the cost of supply; agricultural tariffs cover only about 30 percent. The price/cost ratios for other sectors (commercial, industrial) are much less distorted, with tariffs covering 83–97 percent of the cost of supply.

Consumption levels for residential consumers vary dramatically in Mexico by seasonal tariff zones (figure); as expected, consumption is much higher among customers paying lower tariffs in the more highly subsidized geographic areas (that is, those with higher summer temperatures). Average consumption in the fifth decile of Tariff 1 (least subsidized) is just 97 kWh per month per household, whereas average consumption in the same decile for consumers in Tariff 1F (most subsidized) is 277 kWh per month. The difference between consumption levels in decile 10 is even larger: the largest-volume consumers in Tariff 1 use 270 kWh a month on average compared with 1,240 kWh in Tariff 1F.

Monthly Electricity Consumption by Tariff Category and Consumption Decile

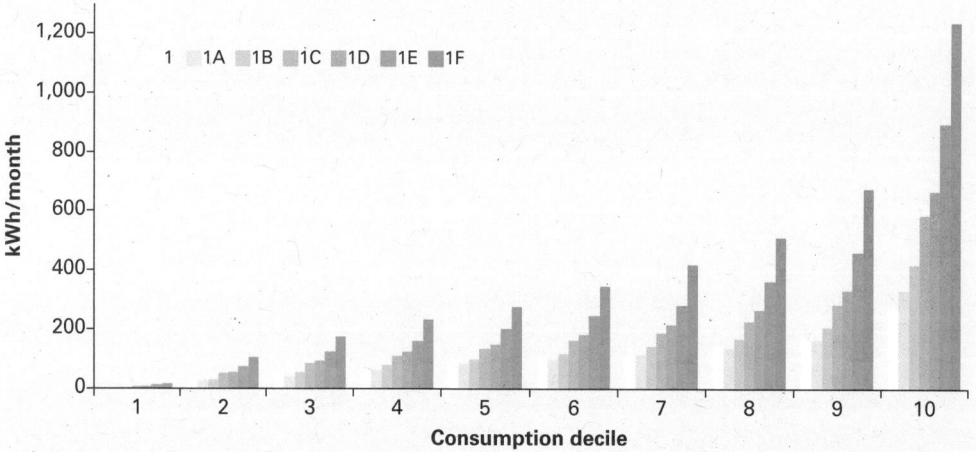

Underpricing of electricity reduces the incentive for customers to take energy-saving measures, such as replacing old equipment and appliances. Elevated demand leads to incremental emissions from power plants, not only of greenhouse gas emissions but also of local pollutants, such as particulates and ozone precursors, which are responsible for the majority of air pollution health impacts. Furthermore, electricity subsidies for irrigation pumping in Mexico have contributed to the overexploitation of groundwater resources in numerous localities.

The bulk of Mexico's electricity subsidies go to the nonpoor. In 2005 the bottom three residential income deciles accounted for about 21 percent of total subsidies, whereas the top three income deciles accounted for 38 percent. By contrast, the pilot program Oportunidades Energéticas has a very progressive distribution of resources across income classes, with nearly 75 percent of energy payments going to the bottom three income deciles.

Source: Komives and others 2009.

rather than life-cycle costs, and the hiring of ESCOs, which are often involved in both the initial audit and project implementation, can be nearly impossible.[10] Mexico also has very restrictive contracting policies at the federal and state levels, which prevent contracts from being awarded for more than one year (because of budget cycles and future obligations), thus limiting efficiency investments.

After studying these issues in 2005, the U.S. Agency for International Development suggested that, rather than seek sweeping changes in procurement and budgeting policies, the government consider implementing a few pilots to test alternative approaches. Simplified ESCO contracts, which could be adjusted to fit one-year contract restrictions with simpler performance verification requirements, could be tested and, based on implementation experience, adjusted and replicated. As these bidding schemes become more accepted, an increased emphasis could be placed on bundling multiple public facilities together in order to scale up investments while reducing transaction costs. This effort could be complemented with a public revolving fund, perhaps on concessional terms, to improve the incentives for energy-efficiency projects to be implemented. This experience would then allow more informed, comprehensive revisions of public policies to be considered.

Conclusions

Mexico's energy-efficiency, cogeneration, and renewable energy interventions in all stationary energy end-use sectors should be an important component of climate change mitigation policy. This has been clearly recognized in the National Strategy on Climate Change. Many of the MEDEC interventions represent accelerated or scaled-up activities that CONUEE, FIDE, and other agencies are already undertaking.

The MEDEC interventions are mainly retrofit/renovations that involve replacement of existing equipment with new and more efficient ones. As Mexico's electricity demand is projected to more than double by 2030, it is important that new equipment meets increasingly stringent energy-performance standards. In this regard, Mexico will benefit from the increasing harmonization of MEPS with the United States and Canada. Mexico would also benefit from a concerted effort to stop the inflow of old and inefficient equipment from the United States. In the rural sector, the dissemination of improved fuelwood cookstoves has significant greenhouse gas mitigation potential, with additional co-benefits.

Energy-efficiency standards for buildings is an area in which Mexico can make significant progress. Doing so requires creating incentives for local governments to adopt and enforce the federal commercial building energy standard and introducing (preferably) mandatory energy-efficiency standards for residential buildings, at least in warm areas with high air-conditioning demand.

There is much potential for ESCOs in advancing Mexico's energy-efficiency agenda, especially in the industrial, commercial, and public sectors. Exploiting this potential will require increased support to developing and piloting ESCO business models that suit Mexico's circumstances. Reducing broad-based residential electricity price subsidies while providing properly targeted support for low-income households, will contribute to improving the incentives for energy conservation and investment in more efficient residential equipment.

Notes

1. This figure does not include energy production and conversion sectors, such as power generation and oil and gas.

2. About 22,000 GWh a year included in industrial electricity consumption are actually attributable to the commercial and services sector, because a number of large nonindustrial buildings, such as hotels, supermarkets, and hospitals, pay an industrial electricity tariff, the basis on which the data are collected. Therefore, industrial electricity use is overestimated and commercial and public services use underestimated by current electricity statistics in Mexico (estimate by Odón de Buen, energy efficiency consultant, 2009).

3. The direct reduction process uses a gas (in Mexico's case, natural gas) to reduce iron ore to produce direct-reduced iron, which can then be fed into electric arc furnaces. Electric arc furnaces account for roughly three-quarters of Mexico's steel output, one of the highest shares in the world (IEA 2007).

4. Overall energy use data are based on IEA (2008a). According to the SENER Energy Information System for 2008 (SENER 2008d), electricity consumption by the commercial and public services sector is 24,300 GWh per year (11 percent of Mexico's total). However, the actual figure for commercial and public service electricity consumption is closer to 46,300 GWh per year, or 21 percent of the total (see note 2 above).

5. The "upstream" benefits of electricity savings in terms of forgone electricity-generation capacity, fuel use, operations and maintenance costs, and emissions were calculated according to the same assumptions of the power sector—that is, that a mix of coal and natural gas generation is displaced. Several teams carried out these analyses: the land-use and bioenergy team developed improved cookstoves and bagasse cogeneration; Odón de Buen (energy efficiency consultant) analyzed street lighting, nonresidential lighting, and nonresidential air conditioning; the electricity team was involved in the two cogeneration interventions; the energy-efficiency team was in charge of the remaining interventions.

6. The federal government is planning large-scale residential refrigeration and lighting programs. It is therefore possible that the proposed MEDEC interventions will actually be part of the baseline.

7. Moreover, in the steel and iron sector, the waste heat fluids are highly corrosive and therefore difficult to handle. Cogeneration projects are basically of two different types of power cycles, topping or bottoming. The *topping cycle* is the most widely applicable in industry, where waste heat from an electrical or mechanical power process is used. A *bottoming cycle* uses the waste heat from a heating process, which is typically supplied to a steam turbine, extracting steam to the heating process and also generating electrical power (Sci-Tech Encyclopedia 1997).

8. This report benefited from substantial research on the health and climate change impacts of adopting improved cookstoves in Mexico (Armendáriz and others 2008; Johnson and others 2008, 2009).

9. Efforts include the energy-efficiency financing study by the U.S. Trade and Development Agency (USTDA) and Nacional Financiera (Nafin), Mexico's largest development bank (USTDA/Nafin); the energy-efficiency financing protocols developed by the Asia-Pacific Economic Cooperation (APEC) and CONAE (APEC/CONAE); the special-purpose financing vehicle for energy efficiency developed by the Energy Sector Management Assistance Program (ESMAP) and the North America Development Bank (NADB) (ESMAP/NADB); the performance and credit risk mechanism developed by the Renewable Energy and Energy Efficiency Partnership, EPS Capital Corporation, and Nafin; and several energy-efficiency/clean energy financing programs developed by FIDE, the Clean Tech Fund, Fondelec Capital Advisors company, NADB, the Japan Bank for International Cooperation, and Nafin.

10. The fact that each ESCO bids on a different project, with different investment needs, energy savings, share of savings to the customer, and other details can make it difficult to evaluate transparency.

CHAPTER 5

Transport

Transport is the largest and fastest-growing sector in Mexico in terms of energy consumption and greenhouse gas emissions. The sector consists of the road, air, rail, and water transport subsectors. It produces about 18 percent of total greenhouse gas emissions in Mexico, with road transport accounting for about 90 percent of energy consumption and CO_2e emissions from the transport sector (SEMARNAT 2007).

Energy use by road transport in Mexico increased more than fourfold between 1973 and 2006, compared with the approximate doubling of energy use by industry and other sectors (IEA 2008a). The country's vehicle fleet nearly tripled in a decade, increasing from 8.3 million vehicles in 1996 to 21.5 million vehicles in 2006.

The import of used vehicles from the United States has been an important factor behind the growth of the vehicle fleet. It has also led to an increase in the average age of the fleet and related problems of low gas mileage and high emissions of criteria pollutants (CO, NO_X, SO_X, and particulates). In 2005 alone, Mexico imported 1.3 million vehicles from the United States that were more than 10 years old (CTS 2009).

Over the next 25 years, Mexico's motorization rate—defined as the number of vehicles per 1,000 people—is projected to continue to increase, following a worldwide trend (figure 5.1). Important factors explaining the increase in motorization in Mexico include the increase in per capita income, the availability of inexpensive vehicles (new and used), and the relatively low cost of transport fuels. Other factors that have contributed to increasing energy use and greenhouse gas emissions from the transport sector are the deteriorating quality of public transportation, the inadequate enforcement of vehicle emission standards, the neglect of transportation needs in urban development plans, and the lack of regulation of freight transport.

Figure 5.1 Motor Vehicle Ownership: Historical Trend and Projected Growth for Selected Countries

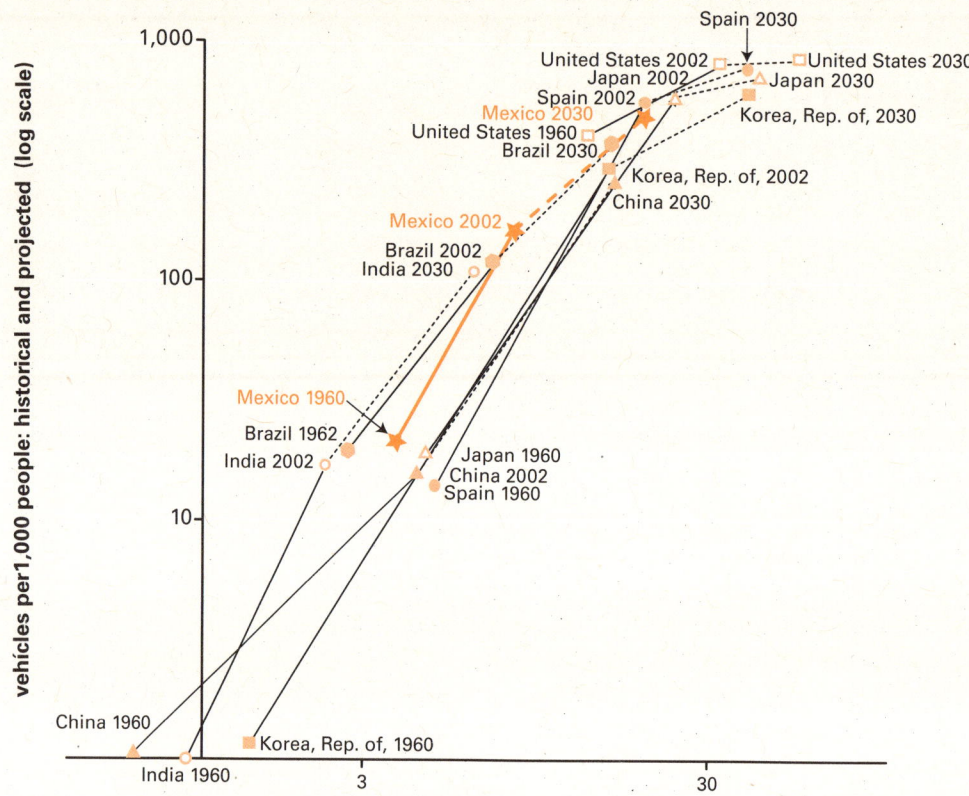

Source: Dargay, Gately, and Sommer 2007.
Note: PPP = purchasing power parity.

One additional factor that contributes to demand for transport fuel is fuel pricing. The prices of the two primary road transport fuels—gasoline and diesel—remained stable or fell over the past 15 years in Mexico (figure 5.2). Fuel prices in Mexico are lower than those of most countries in the Organisation for Economic Co-operation and Development.

The Baseline Scenario

The baseline scenario follows historical trends in Mexico and is consistent with the pattern of motorization growth worldwide. Under this scenario, the national fleet increases from 24 million vehicles in 2008 to a little more than 70 million vehicles in 2030 (figure 5.3). The majority of the increase is for passenger cars, but there is also a large increase in light-duty trucks, buses, and sport utility vehicles (SUVs). Greenhouse gas emissions from the

Figure 5.2 Gasoline and Diesel Fuel Prices in Mexico, 1980–2007

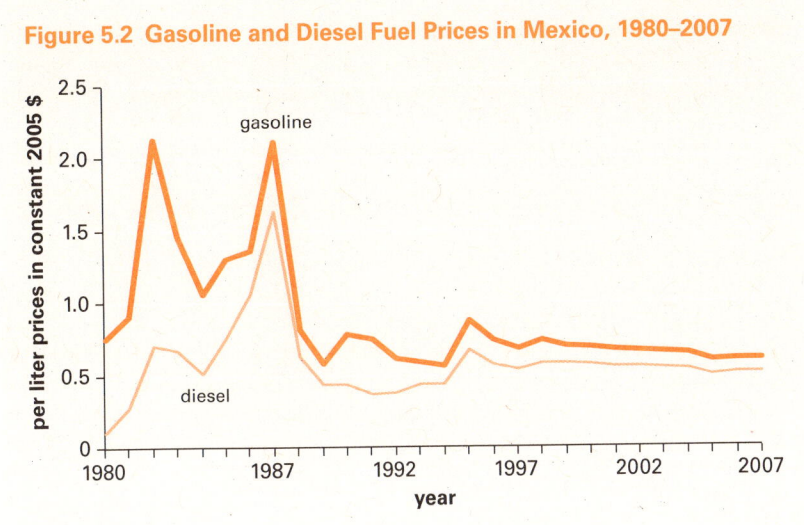

Source: CTS 2009.

Figure 5.3 Transportation Fleet: Historical Trend and Projected Growth under the Baseline Scenario, 1980–2030

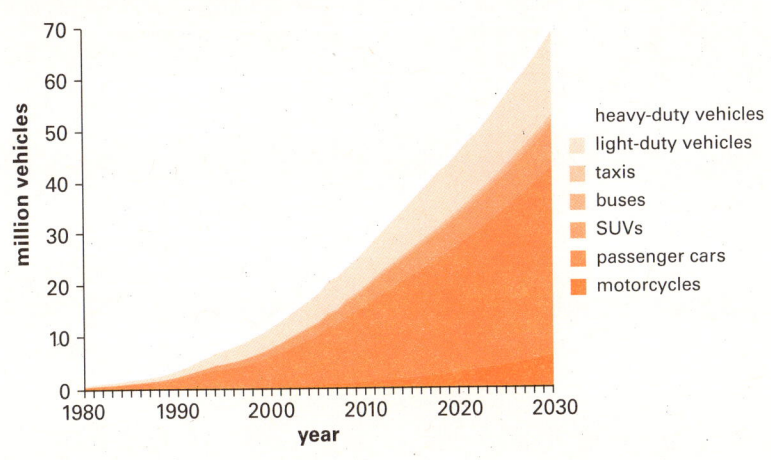

Source: Authors.

transport sector increase from 167 Mt CO_2e in 2008 to more than 347 Mt CO_2e in 2030, with 72 percent of the emissions (and energy consumption) generated by private vehicles (passenger cars, SUVs, and light- and heavy-duty vehicles) (figure 5.4). Total emissions rise from 659 Mt CO_2e in 2008 to 1,137 Mt CO_2e in 2030, with transport's share rising from 25 percent to 31 percent (figure 7.1).

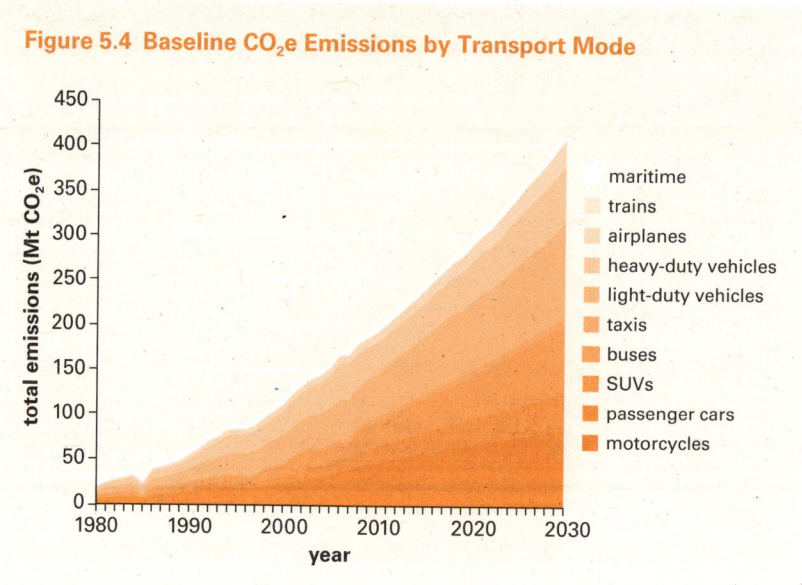

Figure 5.4 Baseline CO₂e Emissions by Transport Mode

Source: Authors.

Figure 5.5 MEDEC Emissions Scenario for Transport

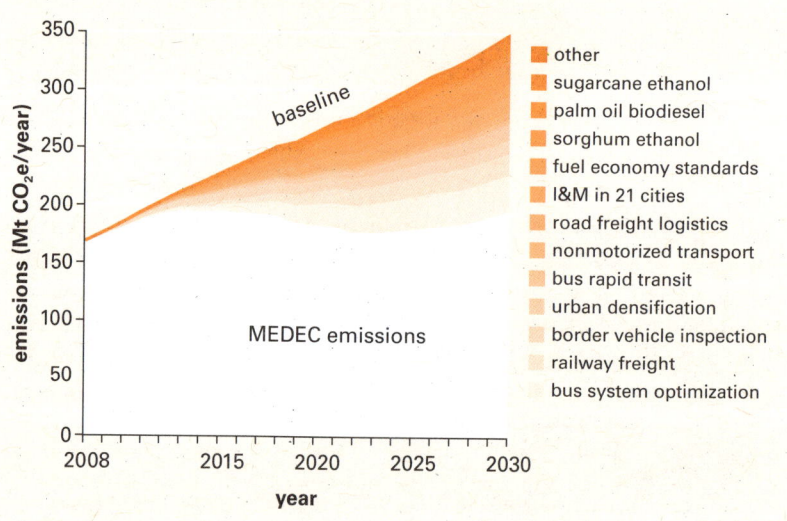

Source: Authors.
Note: I&M = inspection and maintenance. Figure includes all interventions that lead to a reduction in transport sector emissions; this includes those addressed in this chapter as well as the biofuel interventions outlined in chapter 6.

The MEDEC Low-Carbon Scenario

The transport analysis used a programmatic approach to evaluate an integrated set of nine low-carbon interventions.[1] The objective was to identify an aggressive scenario that could dramatically reduce Mexico's transport-related greenhouse gas emissions. The priority areas evaluated in the study include urban land-use, fuels and technology, public transit, nonmotorized transport, travel demand management, and freight transport.

Modal Shift and Urban Development

Urban densification. This intervention seeks to promote a policy for the development and preservation of urban centers, using sustainability criteria that offer conditions of livability (access to work, schools, shops). Urban planning that incorporates increased density makes it possible to reduce the demand for motorized transportation while revitalizing urban centers with mixed land use; recovering the urban landscape; and rebuilding communities by providing equal access to goods and services, education, and maintenance of environmental and urban quality. High-density urban planning imposes growth limits on urban zones, directly affecting the use of vehicles (private and public) and fuel consumption. The cost-benefit analysis considers the changes in infrastructure investment and operation costs (lower in the high-density scenario) and in distances traveled (shorter in high-density areas).

Bus rapid transit. BRT refers to the substitution of minibuses in the main axis routes by rapid mass transit systems of the type introduced in several cities in Mexico (León, Mexico City, and Guadalajara). Systems would be introduced in Mexican cities that currently have more than 750,000 inhabitants. The target of the program is to have 1.5 kilometers per 100,000 inhabitants of BRT lanes by 2030, equivalent to 122 lines of BRT systems, with a total of 1,830 kilometers nationwide. The analysis assesses the mitigation resulting from a fraction of passengers switching from other more polluting means of transport (minibuses as well as passenger cars and taxis) to BRT.

Bus system optimization. This intervention involves the restructuring of the mass transit system's feeder routes by removing redundant vehicles. If complemented by improvements in urban infrastructure (roads, bus stops, traffic signs); public information; traffic monitoring; control; and vehicle improvements, this measure represents an important option for mitigating greenhouse gas emissions in urban public transportation, because the growth of the private vehicle fleet (and related issues of urban sprawl and congestion) has been at least in part the result of inefficient transportation systems.

Nonmotorized transport. Nonmotorized transport is a mobility alternative that gives priority to pedestrians and bicyclists, mostly for short trips. It is an efficient, accessible, nonpolluting means of transportation that is beneficial to health and has recreational value. Formal nonmotorized transport systems are typically used as feeder systems to mass transit systems for longer-distance trips; they should be interconnected with the most important trip destinations (schools, work, shopping centers, tourist sites). Under this scenario, the study quantified a 5 percent national modal share for bicycle trips by 2030. The cost and benefit data are based on studies undertaken in cities that have undertaken effective nonmotorized transport infrastructure programs.

Technologies and Demand Management

Border vehicle inspection. Border vehicle inspection would indirectly regulate the efficiency of used imported vehicles by requiring such vehicles to meet minimum environmental standards. Vehicles that exceed the 2 percent CO (volume) threshold—20 percent of imports in 2006—would be restricted from being licensed in Mexico.

Inspection and maintenance in 21 cities. A program of vehicular use restrictions would be implemented through inspection and maintenance in 21 cities. The objective of the program would be to deter the use of private vehicles and allow the promotion of sustainable mass transit. Within Mexico's current legal framework, the implementation of such a program would lie with state or municipal level authorities; it would be politically difficult to enact it at the federal level. This intervention therefore assumes the adoption of a vehicular inspection and maintenance program similar to the program in place in Mexico City as well as vehicle use restrictions for older vehicles in 21 other metropolitan areas, which would cover about 60 percent of Mexico's total vehicle fleet (without including Mexico City).

Fuel economy standards. This intervention would provide a regulatory incentive to promote more efficient technologies for new vehicles. An energy-efficiency standard based on the weighted average of sales, fuel consumption, and the total number of vehicles manufactured for sale in the country was evaluated for its impact on energy consumption and greenhouse gas emissions. Assuming an increase in vehicle prices as a result of the CAFE–style standard,[2] this measure runs the risk of encouraging sales of used cars, which could reduce fuel economy if implemented in isolation. Therefore, standards for new vehicles should be accompanied by mechanisms that discourage the purchase and ownership of inefficient used vehicles, such as the inspection and maintenance and border inspection interventions outlined above.

Freight

Road freight logistics. This intervention aims to optimize freight transportation by coordinating the operation of heavy-duty vehicles. It includes the creation of freight enterprises or cooperatives, specialized terminals, freight transportation corridors, and information systems. Despite higher fixed costs arising from the companies' infrastructure and management, net costs (and emissions) would be lower, because of the reduction in empty trips.

Railway freight. This intervention would expand the use of the railroad sector from 7.6 percent of all national transported freight in 2007 to 37 percent by 2030. The increase in rail transport would come at the expense of truck freight, although road freight transport would continue to grow in absolute terms, driven by economic growth.

Summary

The analysis of urban transport interventions considered the time savings associated with the reduction in congestion as well as the positive health impacts caused by the reduction in local pollutant emissions (box 5.1). Even without considering these co-benefits, all transport interventions show positive overall cost savings (net benefits) for mitigating emissions (table 5.1).

Other interventions in the transport sector were considered and assessed but ultimately not included in the MEDEC scenario, because they did not

Box 5.1 More Time and Better Health: Co-Benefits of Reducing Emissions in the Transport Sector

In addition to reducing emissions, all of the urban transport interventions examined had significant co-benefits. By reducing the distance traveled by the vehicle fleet, the reduction in congestion leads to time savings. The reduction in local pollutant emissions leads to lower health costs by decreasing the rate of respiratory illness.

These time and health impacts were assessed for all seven of the non-freight MEDEC transportation interventions (figure). The analysis estimates the average time savings likely to result from the interventions, conservatively valuing time at the minimum wage. The health analysis used externality cost factors per liter of fuel burned in urban areas, which were derived from a model that considered estimates of the exposure to local pollutants (PM2.5, NO_x, SO_2, and SO_4) by the affected population. The methodology was adapted from a study by the Instituto Nacional de Ecología (INE 2006), which used exposure response relationships between pollution exposure and health impacts, including cardiovascular mortality, pulmonary mortality, infant respiratory mortality, chronic bronchitis, lost work days, and restricted activity days. Together these co-benefits can be significant for some transportation interventions, providing a major rationale for implementation.

Externality and Time Costs for MEDEC Transport Interventions

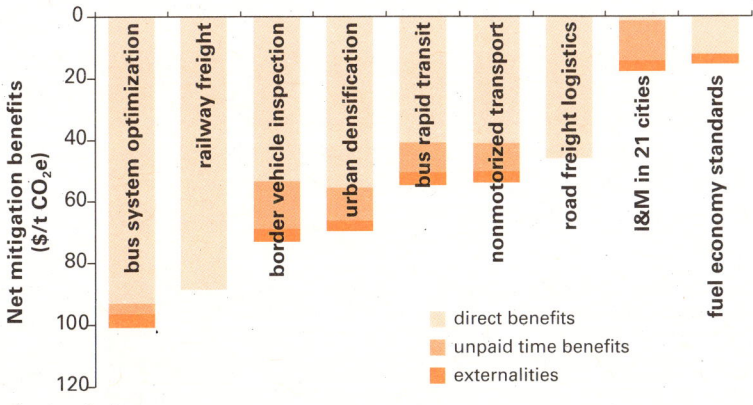

Source: Authors.
Note: I&M = inspection and maintenance.

Table 5.1 Summary of MEDEC Interventions in the Transport Sector

Intervention	Maximum annual emissions reduction (Mt CO$_2$e/year)	Net cost or benefit of mitigation ($/t CO$_2$e)
Modal shift and urban development		
Bus system optimization	31.5	96.6 (benefit)
Urban densification	14.3	66.4 (benefit)
Bus rapid transit	4.2	50.5 (benefit)
Nonmotorized transport	5.8	50.2 (benefit)
Technologies and demand management		
Border vehicle inspection	11.2	69.0 (benefit)
I&M in 21 cities	10.6	14.5 (benefit)
Fuel economy standards	20.1	12.3 (benefit)
Freight		
Road freight logistics	13.8	46.3 (benefit)
Railway freight	19.2	88.7 (benefit)

Source: Authors.
Note: I&M = inspection and maintenance.

meet the MEDEC criteria, because data were not available, or for other reasons. These included the introduction of hybrid vehicles, which have mitigation costs well above the $25/t CO$_2$e threshold; the introduction of diesel vehicles (passenger cars and SUVs), whose mitigation costs were also high; and other travel demand management interventions, such as parking restrictions or congestion charges, on which insufficient information was available. Besides railway freight transport, which was assessed as one of the MEDEC interventions, the redevelopment of railway passenger transport in Mexico is also a promising, although smaller, mitigation option.

Barriers to Mitigating Greenhouse Gas Emissions

Implementation of the aforementioned interventions faces political, financial, and social barriers. An important barrier for the optimization of urban transportation systems is the lack of coordination between agencies working on environment, urban planning, and transport issues, as well as across different levels of governments. The typical result has been an oversupply of low-quality public transport and a lack of overall metropolitan development and mobility planning.

Mass transit interventions also face the challenge of changing the institutional framework and the stakeholders who work in this subsector. In particular, the large number of buses and small concessions for different routes has made it difficult to implement BRT systems or mass transit optimization programs in Mexico. Successful implementation of BRT requires negotiations with route concessionaires who operate along prospective BRT corridors. Demand studies that identify the optimal location for the corridors and technical advice for system planning and operation are also needed.

The most important barrier to vehicular restriction through inspection and maintenance is the lack of enforcement of federal environmental regulations for vehicle emissions, which must be implemented at the state level. As the primary benefit of vehicle inspection programs is on the reduction of local pollutants, the best way to enforce compliance is through public education about health impacts. Vehicle inspection programs can also have an important impact on reducing CO_2e emissions by restricting the use of old vehicles that are both highly polluting and energy inefficient.

Conclusions

Reliance on private vehicles is not a sustainable transport option for Mexico. Although the increase in vehicle ownership in Mexico is probably inevitable, it is possible to substantially reduce vehicle emissions through policies that improve vehicle efficiency, expand and improve public transportation, and optimize the movement of freight. The analysis concludes that all nine transport measures evaluated produce financial and economic savings, as well as yield other benefits, including reduced congestion, pollution, and greenhouse gas emissions.

Because many transport options are interdependent and complementary, it is important that transport issues be addressed in a holistic and programmatic approach rather than as a set of individual measures. Given the historical and future urbanization pattern in Mexico, urban transport and related issues of land-use planning will be a critical determinant of the country's transport energy use and associated emissions. Improving urban transportation will require developing mechanisms that integrate public transportation with urban planning and development efforts by the federal, state, and municipal governments. Although low-carbon development can be an additional consideration, the underlying drivers of sustainable transport policies will be efficient, safe, and clean access to school, work, shopping, and neighborhoods.

Notes

1. The analysis of all transport sector interventions was carried out by the transport team.

2. The standard evaluated for Mexico is similar to the vehicle efficiency standard for new vehicles in the United States known as the corporate average fuel economy (CAFE) standard.

CHAPTER 6

Agriculture and Forestry

The agriculture and forestry sector generated about 135 Mt CO_2e of greenhouse gas emissions in 2002 (PECC 2009), accounting for 21 percent of Mexico's total emissions. Two-thirds of the emissions were generated by the forestry subsector; the remainder came from agriculture and livestock. This chapter examines a set of low-carbon interventions in the rural sector that reduces emissions from agriculture and forestry. It also presents several biomass energy interventions that use crops, crop residues, and sustainable fuelwood that reduce emissions in other sectors (transport, power, industry, and residential) by replacing fossil fuel energy.

Mexico has a surface of 198 million hectares, of which 15 percent is used for agricultural crops and 58 percent is used for some form of grazing. Forests cover 67 million hectares, or 34 percent of the country. In 2006 the agricultural, forestry, and fishing sectors accounted for 5.4 percent of GDP (SAGARPA 2007a).

The forestry subsector has been identified as one of the key areas for greenhouse gas mitigation in Mexico (Masera, Cerón, and Ordóñez 2001), in terms of both avoiding emissions through such actions as reducing deforestation and capturing carbon in forest soils and biomass.

There are fewer cost-effective measures for reducing greenhouse gas emissions in the agricultural sector. Minimum-tillage crop production appears to be a promising technology for Mexico to reduce energy use and aid in soil carbon sequestration. The production of liquid biofuels faces financial and economic barriers, and more research and development needs to be conducted on other low-carbon measures in the agricultural and livestock sectors.

Bioenergy produced by both forestry and agriculture systems represents 8 percent of the primary energy consumption in Mexico (408 petajoules), mainly from the consumption of fuelwood (78 percent) and sugarcane bagasse (22 percent). An estimated 25 million people in rural areas of Mexico—one-fourth of Mexican households—use fuelwood, mainly for cooking.[1] Fuelwood is also used in many small industries, such as pottery and brick-making. Sugarcane bagasse is the basic fuel used in sugar mills. Modern bioenergy has great potential for reducing greenhouse gas emissions and contributing to medium- and long-term energy diversification in Mexico.

The Baseline Scenario

Under the baseline scenario, emissions from the agriculture and forestry sector decrease slightly, from about 100 Mt CO_2e a year in 2008 to 87 Mt CO_2e in 2030. Agriculture and livestock accounted for 7 percent of greenhouse gas emissions in Mexico in 2002 (SEMARNAT and INE 2006a); the baseline scenario assumes that these emissions remain at roughly the same levels in absolute terms. The forestry subsector contributes about 14 percent of greenhouse gas emissions, mostly because of deforestation. The baseline assumes that greenhouse gas emissions from the forestry sector remain constant in absolute terms but also that, based on current reforestation and afforestation trends, net forestry emissions decline slightly over the coming decades.

Historically, three patterns of deforestation have been observed in Mexico: (a) clearing of temperate coniferous, tropical, and subtropical forests for subsistence agriculture and cattle grazing; (b) deforestation in tropical forests associated with the settling of land under the agrarian reform; and (c) land clearing for commercial large-scale cattle ranching and farming. Deforestation by small farmers has been decreasing over the past 20 years because of urban migration and because government-supported land settlement has officially ended.[2] The clearing of forests for large-scale agriculture may be more or less intense in the future depending on market conditions and government land policy.

The MEDEC Low-Carbon Scenario

The study identified and evaluated mitigation interventions within the forestry, agriculture and livestock, and bioenergy subsectors.[3] Twelve interventions met the criteria for emissions reduction, cost less than \$25/t CO_2e, and were judged to be feasible to implement based on existing programs and pilots in Mexico and other countries. The potential for all agriculture and forestry sector interventions was assessed by means of a geographic information system that included the main features of Mexico's territory (figure 6.1). All interventions comply with designated land-use regulations, including adequate set-aside areas for conservation, and avoid competition between food and bioenergy production.

Forestry

The MEDEC forestry interventions include a range of biomass production and forest management programs. Interventions in this subsector can be

Figure 6.1 Geographic Distribution of Agriculture and Forestry Sector Interventions

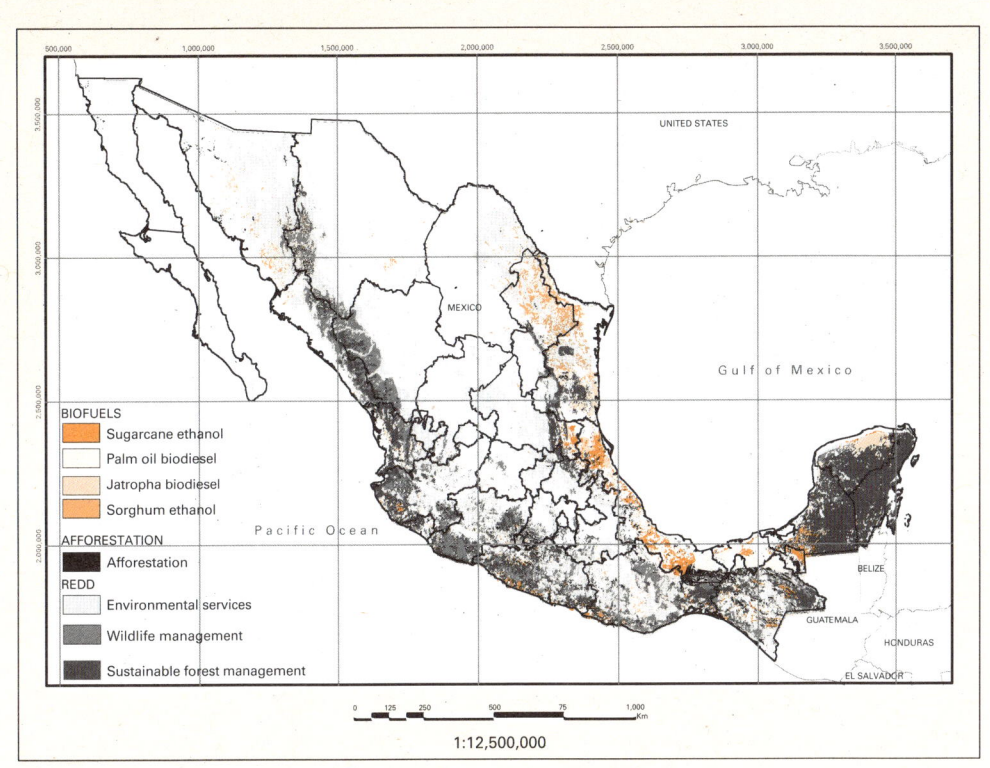

Sources: Ghilardi and Guerrero 2009, based on REMBIO 2008; INEGI 1995, 2000, 2002. Created in ArcGIS 9.2 using ArcMap.

Note: Sustainable forest management includes all interventions that involve a productive use of biomass (biomass electricity, fuelwood co-firing, charcoal production, and forest management). Areas suitable for reforestation and restoration or for zero-tillage maize are not included. The area depicted for afforestation assumes eucalyptus plantations. Jatropha biodiesel, an intervention not included in the MEDEC scenario because of its high net cost of mitigation, is included.

divided into those that reduce emissions from deforestation and forest degradation (REDD)[4] and those that contribute to the reforestation or afforestation of deforested or degraded land (table 6.1). REDD interventions can be divided into those that entail some form of productive use of the woody biomass and those that do not. When the woody biomass is used as a fuel (biomass electricity, fuelwood co-firing, and charcoal production interventions), it displaces the use of fossil fuels. Those interventions therefore reduce emissions through both a REDD and a bioenergy effect. Together the six REDD interventions would involve the management, protection, or both of 65 million hectares of forests, resulting in a zero rate of deforestation and degradation in 2030.[5]

Biomass electricity. Biomass electricity entails the generation of electricity from fuelwood produced in sustainably managed forests. It is assumed that timber, which represents 30 percent of wood production, is sold for other purposes and that sustainable forest thinning and logging residues are used as fuelwood. Sustainable forest management would be accompanied by measures to stop deforestation and forest degradation. Two hundred small power plants (with a capacity of 25 MW each) would be built in regions with native forests. This labor-intensive intervention could create about 200,000 jobs throughout the country. Although there is no experience with this kind of generation technology in Mexico, its use is widespread in other countries, including Austria, Sweden, and the United States.

Fuelwood co-firing retrofitting. Fuelwood co-firing, which combines up to 20 percent wood with fossil fuels, uses fuelwood produced under the same circumstances as in biomass electricity that is then mixed with coal to generate electricity. Of the three coal-fired power plants in Mexico, the 2,100 MW Petacalco plant (in Guerrero state) is the only one located near forests that can provide an adequate fuelwood supply. The intervention is therefore limited to this plant and involves retrofitting the power plant for handling fuelwood and mixing it with coal.

Charcoal production. About 0.6 million tons of charcoal are produced each year in Mexico to meet the needs of the residential and commercial sectors. This intervention increases charcoal production 13-fold to meet increasing urban demands and replace 75 percent of coke demand in industry. It also replaces traditional earthern kilns with more efficient brick kilns. It is assumed that efficient charcoal kilns would supply 70 percent of urban charcoal consumption by 2030 and 100 percent of industrial demand. There are currently no specific government programs for the implementation of efficient brick charcoal kilns. The technology and practice is widespread internationally, however. Like the previous two interventions, charcoal production assumes that sustainable forest management practices reduce deforestation and forest degradation.

Forest management. Forest management is the last of the four interventions that reduce deforestation and forest degradation through the sustainable production of woody biomass. Unlike the previous three interventions, which use biomass as a fuel and therefore substitute for the fossil fuels, in this intervention woody biomass is used as timber or for other nonenergy purposes.

Wildlife management. Wildlife management would involve the scaling up of activities and experiences of a current program of the federal government that provides certification for wildlife management units (known by their acronym in Spanish, UMAs). It is assumed that the income from wildlife management (mainly in the form of hunting permits) would enable UMAs to reduce deforestation and forest degradation.

Payment for environmental services. Payment for environmental services would expand a current government program that provides direct cash payments to forest owners in exchange for forest protection. It is assumed that the payment would be equal to the opportunity cost of using the land for other purposes and that it would enable the owners to put in place mechanisms to reduce deforestation and degradation.

The first six interventions aim to reduce deforestation and forest degradation through the sustainable production of biomass or other mechanisms. Two other interventions, afforestation (commercial tree plantations) and reforestation and restoration, seek to restore forests in areas that have already been deforested.

Afforestation. This intervention entails the planting of eucalyptus and pine species on 1.5 million hectares of land for the production of marketable timber for sawmills, paper mills, poles, and fuelwood. The survival rate for trees planted on these plantations is assumed to achieve the observed rate over the past several years of 50 percent. It is also assumed that 50 percent of the carbon content of each harvest is emitted to the atmosphere.

Reforestation and restoration. This intervention involves the planting of native species in areas in which native vegetation has been cleared. Unlike afforestation, reforestation and restoration does not assume any productive utilization of forest products. Whereas afforestation is assumed to use high-quality soils, reforestation and restoration use lower-grade soils (with lower opportunity costs).

Agriculture

The agriculture subsector includes changes in maize production practices, and the production of biofuels. Maize has been the most important crop in Mexico since pre-Columbian times. SIACON (2007) reports that some 8.2 million hectares were sown with maize in 2006, equivalent to 38 percent of Mexico's total planted area.

Zero-tillage maize. This intervention involves an increase in the sequestration of carbon in the soil (as well as a minor reduction in diesel consumption). Zero-tillage is defined as the tillage system that keeps at least 30 percent of the surface covered with harvest residues, cover, or litter after sowing.[6] The intervention assumes that the maize-planted area under zero-tillage increases from 0.5 million hectares in 2008 to 3 million hectares in 2030, reaching 50 percent of the commercial maize cropping area. The accumulation and decomposition of plant residues leads to an increase in organic carbon soil sequestration. The reduction of diesel consumption by tractors also reduces emissions slightly.

The liquid biofuel category includes current-generation ethanol and biodiesel technologies that substitute for gasoline and diesel from petroleum. For all of the biofuel interventions, it is assumed that the land required for feedstock production comes from pastures and grasslands and that land cannot be converted from other crops, forests, or protected lands. Some level of indirect competition is, however, impossible to avoid (for example, the displacement of pasture land may increase pasture prices and lead to more agricultural land being used for pasture).

Sugarcane ethanol. This intervention involves the installation of 97 ethanol plants, each producing 170 million liters a year. Each plant would require the production of sugarcane from about 30,000 hectares. The intervention assumes that the use of bagasse would allow the plants to be self-sufficient in energy and to sell surplus electricity to the grid. This intervention would reduce greenhouse gas emissions by displacing gasoline use by ethanol in transport and other fossil fuels by bagasse in the electricity sector.

Sorghum ethanol. This intervention involves the construction of 19 ethanol plants of 165 million liters per year per plant. Each ethanol plant would require the production of sorghum from about 160,000 hectares of land. An important source of revenue in this intervention comes from the sale of dried distillers grains, a by-product of ethanol production.

Palm oil biodiesel. This alternative entails the installation of 21 processing plants with a production capacity of about 34,000 tons of biodiesel per year per plant. Each plant requires about 10,000 hectares of palm planta-tions. Revenues are generated from biodiesel production and the sale of palm oil cake, which can be used as cattle feed.

Summary

Successful implementation of all agriculture and forestry measures would mitigate about 1,700 Mt CO_2e between 2008 and 2030. The six REDD interventions have a combined reduction potential of 1,120 Mt CO_2e, or two-thirds of sectoral emission reductions. Other interventions with high greenhouse gas mitigation potential are reforestation and restoration (10 percent), afforestation (9 percent), and sugarcane-based ethanol (9 per-cent). Together these nine alternatives account for 94 percent of mitigation potential in the sector.

The land use, land-use change, and forestry (LULUCF) emissions reduc-tions (63 percent from reduced emissions, 37 percent from carbon capture) of the 12 agriculture and forestry interventions would amount to 927 Mt of CO_2e, accounting for 54 percent of the total impact of these interven-tions. The remaining 46 percent of emissions reductions would take place in other sectors, through the substitution of fossil fuels by bioenergy in the electricity, industrial, and transport sectors. The MEDEC scenario implies that LULUCF emissions in Mexico would become negative in year 2030—that is, Mexico would become a net sink in terms of LULUCF (figure 6.2).

All of the forestry interventions have large reduction potential. Their reduction costs range from a net cost of $18/t CO_2e to a net benefit of $20/t CO_2e (table 6.1). The REDD and reforestation projects have significant

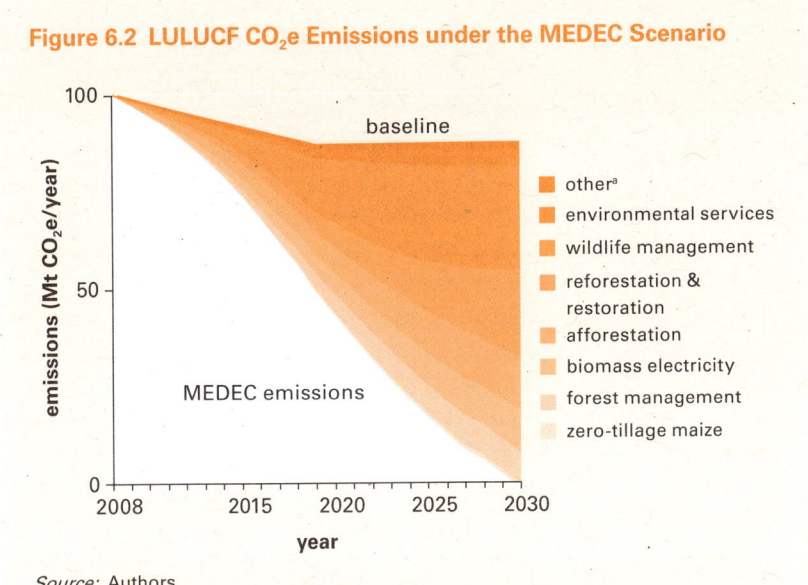

Figure 6.2 LULUCF CO₂e Emissions under the MEDEC Scenario

Source: Authors.

a. "Other" includes charcoal production and fuelwood co-firing, which have a small impact in the reduction of LULUCF emissions. Note that many of the agriculture and forestry interventions produce biomass that substitutes for fossil fuel use in other sectors, including electricity (biomass), transport (biofuels), and heat applications, and are thus shown in other sector emission graphs.

environmental benefits, which were not included in the economic analysis. These co-benefits should be considered (they are discussed in chapter 7). In terms of economic benefits per ton of CO_2e reduced, the most efficient interventions are charcoal production ($20/t) and zero-tillage maize ($15/t).

Other interventions in the agriculture and forestry sector were considered and assessed but ultimately not included in the MEDEC scenario, because they did not meet the MEDEC criteria, because data were not available, or for other reasons. Methane from livestock production, which can be reduced using biodigestors, is a major component of greenhouse gas emissions. Too little information was available on biodigestors for pig or dairy farms, however, and their mitigation potential appeared relatively low. Several crops for biofuels were considered, but adequate data were available for only four crops—the three assessed above plus jatropha, for the production of biodiesel. Jatropha was not included because of its high mitigation costs. Several technologies for generating electricity from biomass were considered, including gasification in several scales. A standard boiler and vapor turbine was finally chosen, for economic reasons.

Barriers to Mitigating Greenhouse Gas Emissions

Although the Mexican government has increased budgets and established new programs in the forestry subsector over the past few years, significant barriers to the implementation of forestry activities remain. Reforestation

Table 6.1 Summary of MEDEC Interventions in the Agriculture and Forestry Sector

Intervention	Surface area (million hectares)	Maximum annual emissions reduction (Mt CO_2e/year)	Net cost or benefit of mitigation ($/t CO_2e)
Agriculture			
Zero-tillage maize (best practice)	2.5	2.2	15.3 (benefit)
Biofuel production			
Sugarcane ethanol	1.5	16.8	11.3 (cost)
Sorghum ethanol	3.2	5.1	5.3 (cost)
Palm oil biodiesel	0.2	2.4	6.4 (cost)
Forestry			
REDD			
With productive use of biomass			
Biomass electricity	11.4	35.1	2.4 (benefit)
Fuelwood co-firing	0.6	2.4	7.3 (cost)
Charcoal production	9.0	22.6	19.6 (benefit)
Forest management	9.0	7.8	12.7 (benefit)
Without productive use of biomass			
Wildlife management	30.0	27.0	17.8 (cost)
Payment for environmental services	5.0	4.4	18.1 (cost)
Reforestation/afforestation			
Reforestation and restoration	4.5	22.4	9.3 (cost)
Afforestation	1.6	13.8	8.4 (cost)

Source: Authors.

and restoration programs could achieve greater success with selected and certified sources of seeds and improved quality of seedlings, training for landowners, and better selection of planting sites. Management of native forests could be greatly improved through closer supervision by forest services; control of illegal logging, fires, and pests; and improved thinning practices. Most of these issues could be addressed through capacity building at all levels, including training programs on seed collection and nursery and forest management, which are among the most urgent measures required. As most forests in Mexico are under some form of community ownership, the implementation of all forestry interventions involves the design of adequate institutional frameworks for community participation.

Charcoal production would encounter some barriers to implementation. These include the lack of a dedicated government program, cultural resistance to the adoption of new production technologies, the need for training and technical assistance to ensure proper use and maintenance of the new technology, lack of capital to invest in kilns and equipment, and the shortage of qualified and certified kiln builders.

The zero-tillage maize farming system is used in Mexico, but there are a number of barriers to its wider implementation. Most farmers are not familiar with it; there is not a well-developed market and market-support structure for associated agricultural services, such as spraying and direct sowing; and it runs counter to the traditional use of maize stubble as forage for cattle.

Conclusions

The interventions in forestry account for almost three-fourths of the mitigation potential in the agriculture and forestry sector, yielding among the largest mitigation gains in Mexico. The analysis does not consider the environmental benefits (such as biodiversity conservation) associated with maintaining and increasing forest cover. The successful implementation of most forestry subsector interventions depends on changes in forest management, public funding, and the development of a market for sustainable forest products. Climate change considerations could provide additional incentives for forestry programs in Mexico. The estimated cost to achieve REDD through the payment of environmental services is about $18/t CO_2e. In contrast, forest management interventions for bioenergy or other purposes, which also produce REDD benefits, have net benefits rather than costs.

Bioenergy has significant potential for reducing emissions at low costs. The lowest-cost intervention in the agriculture and forestry sector is charcoal production; the highest annual mitigation is achieved by biomass electricity.

Liquid biofuels interventions other than sugarcane ethanol were estimated to have limited reduction potential without impinging on land used for food crops, forests, or conservation lands. (There was an explicit assumption not to include lands for biofuels that are currently being used for other crops, forests, or other conservation purposes; in practice, it is difficult to control land conversion if there are profitable uses for the land.) Mexican production costs for sugarcane are significantly above world levels, requiring domestic subsidies for sugar producers. Unless production costs can be dramatically reduced, Mexican ethanol will not be competitive with ethanol produced in other countries.

If all the agriculture and forestry interventions were implemented, the sector could provide about one-third of total national emissions reductions over the coming two decades. About two-thirds of the reduction could be achieved at net costs of less than $10/t CO_2e.

Notes

1. An intervention that addresses improved biomass cookstoves is discussed in chapter 4.

2. Land reform in Mexico, which began in the 1930s and continued through 1992, provided more than 100 million hectares—almost half of the national territory

and some two-thirds of the country's total rural property—to rural Mexicans, who formed 30,000 *ejidos* (cooperatives) and communities. The land transfer included certain restrictions, such as an obligation to actively cultivate the land and a prohibition on the sale or rental of the land, in addition to restrictions on intergenerational land transfer. Some of the state-driven land colonization projects were disappointing: vast forest areas were cleared for agricultural settlements that never achieved their intended production levels. Forest and natural ecosystems were cleared not only for agricultural purposes but also for pasture and for tourism development. These projects favored vested interests and paid little attention to environmental consequences. Unlike elsewhere in Latin America, land distribution in Mexico contributed to social stability during the 1970s and 1980s. The cost of Mexican social peace was paid for with the natural capital of the lowland tropics, however. In 1992 the agrarian legal framework was updated and a series of legal and policy reforms—the National Certification Program of Ejido Rights and Urban Lots [PROCEDE]—was introduced, including a program of land rights regularization targeting the "social sector." Among other things, the program authorized *ejidos* to form joint ventures with private companies, lifted the prohibition on land rental, and authorized land sales with some restrictions intended to keep the plot within the hands of the local community (de Dinechin and Larson 2007).

3. All interventions in this chapter were analyzed by the land-use and bioenergy team. The electricity sector team participated in biomass electricity and in fuelwood co-firing.

4. For the purposes of this study, *deforestation* is defined as the change from forest to any other category of land use. This study also assumes that all of the above-ground biomass of the forest is converted into CO_2e. By contrast, *land degradation* is assumed to result in only a partial loss of the forest biomass.

5. Since 2001 the government has designated an ever-increasing budget to the forestry sector, and reduction of deforestation and forest degradation are key components of the National Forest Strategy for 2002–25 (CONAFOR 2001). In 2007 the various support programs for forest development were united into a single program known as Proarbol. The program includes direct transfer payments to landowners, through various subprograms designed to conserve forests, restore degraded areas, and reforest land, including through financial assistance to communities for forest fire control, pest management, and the introduction of efficient fuelwood stoves in rural areas. The MEDEC scenario assumes that this program would be continued and expanded.

6. Zero- or minimum tillage has been implemented in Mexico since the late 1970s, under scientific guidance. Its benefits include less soil erosion; higher moisture retention; lower soil compaction; lower energy consumption; improved physical, chemical, and biological properties of the soil; reduction of weeds and absence of new weeds; reduced production costs; greater biological activity in the soil; better development of crop roots; and reduction of water deficiency (Navarro 2000; Pitty 1997; Rojas, Mora, and Rodríguez 2002; SAGARPA 2007a).

CHAPTER 7

A Low-Carbon Scenario for Mexico

This chapter presents the aggregate results of the sectoral interventions evaluated under MEDEC, which are used as inputs to an alternative emissions modeling scenario for 2030. This chapter also compares the net costs (benefits) of the low-carbon interventions across sectors in the form of a marginal abatement cost curve. The chapter concludes by presenting the results from a dynamic computable general equilibrium model used to examine the potential impact of the MEDEC interventions on the Mexican economy.

The Carbon Path under the Baseline Scenario

To generate a low-carbon scenario for Mexico, it is necessary to first assess what would happen under the baseline case with no consideration for climate change and assuming an effective carbon price of zero. For this scenario, the study used the LEAP (Long-range Energy Alternatives Planning) model to account for emissions from energy production and consumption activities.[1] Emissions from activities not associated with energy, such as industrial processes and land-use, were modeled separately.

The baseline scenario is based on macroeconomic assumptions that are consistent with those of the government of Mexico, including average annual GDP growth of 3.6 percent,[2] average annual population growth of 0.6 percent, and a set of fuel prices that correspond to a West Texas Intermediate oil price of about $53 per barrel in 2009, which increases slightly in real terms over the period of analysis to 2030. The baseline scenario takes into account both historical trends and the impact of sector policies and programs that are already under implementation (table 7.1).

Based on these assumptions, the baseline scenario estimates that total greenhouse gas emissions in Mexico will grow from 659 Mt CO_2e in 2008

Table 7.1 Key Assumptions and Indicators for Baseline Scenario

Parameter	2008	2030	Assumptions and trends
Population	106.7 million	120.9 million	0.6% annual growth
Urbanization	77%	85%	Official projections
GDP	$734 billion	$1.599 trillion	3.6% annual growth
CO_2e emissions			
Electric power	142 Mt (22%)	322 Mt (28%)	Current growth in power generation is mainly based on imported natural gas; baseline scenario foresees a continued increase but a slowdown in new natural gas–based capacity and an increase in the contribution of coal (mostly imported) in the power sector energy mix
End-use fuel consumption for heat	107 Mt (16%)	160 Mt (14%)	Fuel consumption in all energy end-use sectors except transport is expected to grow at below-GDP growth rates
Transport	167 Mt (25%)	347 Mt (30%)	Increased ownership and use of private cars due to income growth, urban sprawl, and the availability of cheap second-hand cars imported from the United States
Land use	100 Mt (15%)	87 Mt (8%)	A slowdown in deforestation rate is assumed
Waste and industrial processes	143 Mt (22%)	221 Mt (19%)	Increased raw materials consumption and waste production due to continued income growth and urbanization
Total	659 Mt	1,137 Mt	

Source: Authors.

Figure 7.1 Greenhouse Gas Emissions under the Baseline Scenario, by Source

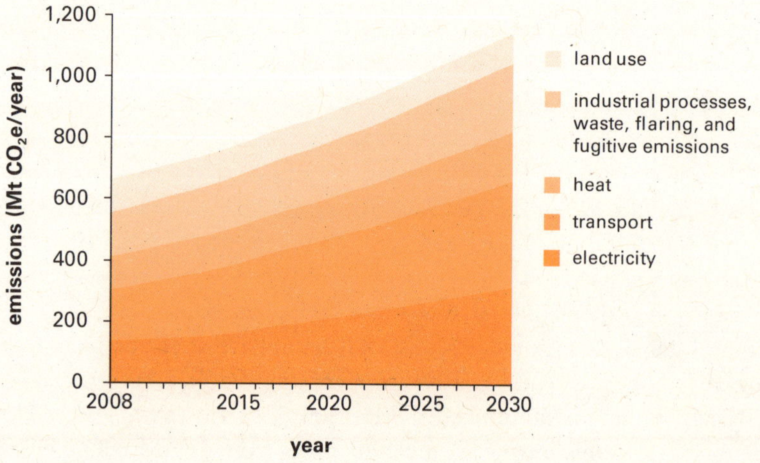

Source: Authors.

to 1,137 Mt in 2030 (figure 7.1). Although this is a substantial increase in total emissions, it reflects a reduction in GDP carbon intensity, from 0.98 kg CO_2e to 0.74 kg CO_2e per dollar. However, per capita carbon emissions would increase from 6.75 t CO_2e to 9.84 t CO_2e, reflecting to a large extent the effect of rising income on energy and materials consumption.[3]

Most of the emissions growth under the baseline takes place in the two sectors that are already the largest contributors, transport and electric power. The share of the transport sector in total emissions is projected to increase from 21 percent in 2008 to 27 percent in 2030; emissions by the power sector are projected to grow from 18 percent to 24 percent of total emissions. Land-use emissions are projected to decrease in absolute terms from 100 Mt to 87 Mt a year, following the historical trend in Mexico of a reduction of emissions from deforestation.

The MEDEC Alternative Low-Carbon Path

The alternative MEDEC scenario is built on the same macroeconomic assumptions as the baseline for GDP growth, population growth, and the rate of urbanization. The key objective of the MEDEC low-carbon scenario is to achieve a similar level of economic growth with a significantly smaller carbon footprint. This is achieved by pursuing cost-effective low-carbon interventions through policies and investments. Forty interventions from five sectors (see table 1.1) are included in the MEDEC scenario, all of which meet the criteria outlined in the evaluation methodology described in box 1.1. The MEDEC scenario reflects the emissions reductions and corresponding implications of implementing only these 40 interventions.

Under the MEDEC scenario, sector policies are assumed to be adopted that maximize the benefits of no-regret low-carbon interventions; to that extent, greenhouse gas emission mitigation becomes an explicit policy objective (table 7.2). In the electric power sector, for example, investments and policies that promote low-carbon alternatives such as cogeneration, wind, geothermal, and hydropower would be pursued. In oil and gas, the objective would be to improve the efficiency of Pemex facilities and to reduce the leakage of gas in distribution and storage. In the energy end-use sectors, investments and policies would mainly scale up and accelerate ongoing initiatives by the government and the private sector, building on previous successes. In transport, investments and policy measures would be pursued to increase the modal share of public transport and other alternatives to private vehicles in urban areas, improve vehicle fleet efficiency, and optimize the movement of freight. In agriculture and forestry, the priority would be to strengthen programs for reforestation and afforestation, reduce deforestation and degradation, and promote the use of sustainable biomass energy.

Implementation of the MEDEC interventions would stabilize Mexico's greenhouse gas emissions at roughly 2008 levels over the period to 2030,

Table 7.2 Results and Key Sector Developments under the MEDEC Scenario

Sector	Cumulative GHG emissions reduction 2008–30 (Mt CO_2e)	Emissions reduction achieved in 2030 (Mt CO_2e/year)	Key sector developments compared with baseline scenario
Electric power	876 (17%)	91	Reduced fossil fuel consumption in power generation by increasing the utilization of low-carbon renewable energy technologies
Oil and gas	435 (8%)	30	Reductions in gas leakage in natural gas transportation, cogeneration in Pemex, and refinery efficiency
Energy end-use	857 (16%)	63	Reduced electricity demand by tightening minimum energy performance standards and accelerating programs to replace inefficient appliances, lights, and industrial motors; reduced fuel demand by scaling up solar water heating in households and cogeneration in industries; lower CO_2 and other emissions by accelerating the dissemination of improved fuelwood cookstoves
Transport	1,422 (27%)	131	Lower fossil fuel demand by promoting higher density urban growth, efficient mass transit, nonmotorized transport, vehicle fleet efficiency, and improved logistics and increased use of rail for freight transport
Agriculture and forestry	1,706 (32%)	162	Expanded programs for reducing deforestation and degradation, reforestation and afforestation, forest management, and sustainable fuelwood and biomass energy production
Total	5,296 (100%)	477	Steady economic growth without increasing Mexico's carbon footprint and with significant co-benefits

Source: Authors.
Note: GHG = greenhouse gas.

reducing CO_2e emissions by about 477 Mt relative to the baseline (figure 7.2). Under the MEDEC low-carbon scenario, it is possible for income in Mexico to grow steadily while maintaining carbon emissions at roughly the same level.[4]

The three sectors that currently account for the majority of greenhouse gas emissions also have the greatest potential for cost-effective emissions reduction under the low-carbon scenario.[5] Of total cumulative greenhouse gas emissions reduction in the MEDEC scenario, 27 percent come from transport, 17 percent from electricity generation, and 32 percent from agriculture and forestry. Measures in the energy end-use sectors, mostly resulting in reduced electricity demand, account for about 16 percent of the cumulative greenhouse gas emissions reduction. The oil and gas industry contributes the remaining 8 percent (table 7.2).[6]

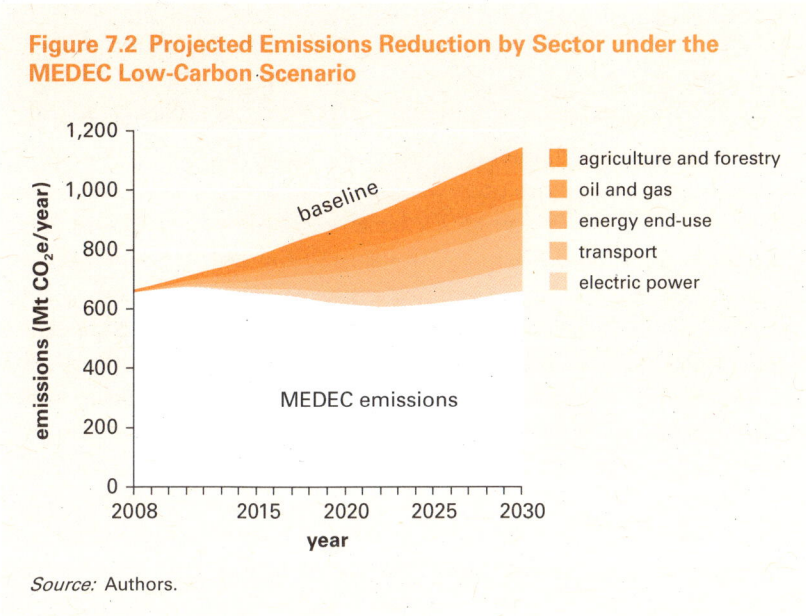

Figure 7.2 Projected Emissions Reduction by Sector under the MEDEC Low-Carbon Scenario

- agriculture and forestry
- oil and gas
- energy end-use
- transport
- electric power

baseline

MEDEC emissions

Source: Authors.

The Net Costs (Benefits) of Emissions Reduction

One of the main objectives of the MEDEC study is to quantify the benefits and costs of potential greenhouse gas mitigation options using a consistent methodology. In this way, MEDEC interventions from different sectors can be compared based on a robust economic analysis (see box 1.1 and annex B). The MEDEC study is not a comprehensive assessment of possible mitigation interventions in Mexico.[7] Other promising low-carbon interventions could be subjected to a similar type of analysis to compare their reduction potential, costs, and investment requirements with the 40 MEDEC interventions.

The study took a two-part approach to determining the net costs (benefits) of the low-carbon interventions. In the first step, the analysis is limited to measurable financial and economic costs and benefits—such as the level of new investment, avoided investments, operating costs, and a stream of benefits, such as the value of energy savings—for all of the important stakeholders. In a second step, the externality costs and benefits are identified and evaluated. The approach, similar to that used in a typical World Bank financial and economic appraisal of an investment project, would include such results as profitability, income generation, and an assessment of the social and environmental externalities (both positive and negative).

The quantitative environmental externality analysis undertaken for the MEDEC study was limited to the health impacts associated with reducing local air pollution (primarily for transport, household fuel use, and electric power generation). Because comparable data were not available for most interventions, and because only air pollution externalities were assessed, the environmental externality results are not included in the marginal abatement cost curve; these results are reported separately. Other costs and

benefits that were not included in the economic analysis for MEDEC interventions include transaction costs, such as the political cost of passing and
implementing new legislation, and other more tangible but also difficult to
quantify costs, such as the need to inform consumers, develop public or
private institutions, and build new businesses and markets.

Many interventions with positive economic benefits are not being implemented and are not likely to be implemented until key barriers are overcome. Some of the major barriers inhibiting the implementation of MEDEC
interventions were discussed in the sector analysis chapters (chapters 2–6).
A number of the broader policy and investment barriers are discussed in
chapter 8. Additional work is needed to assess the institutional, behavioral,
and other barriers that inhibit low-carbon interventions from being implemented and how such barriers can be overcome.

The results of the economic evaluation are summarized in the combined
marginal abatement cost curve (figure 7.3). Interventions in the upper half
of the curve have net incremental costs; interventions in the lower half have
net incremental benefits. The area of each bar represents the total net cost
(benefit) of a MEDEC intervention. Some bars are either too narrow (small
emissions abatement, such as efficient street lighting) or too small (low unit
net costs or benefits, such as biogas) to be visible.

Figure 7.3 Marginal Abatement Cost Curve

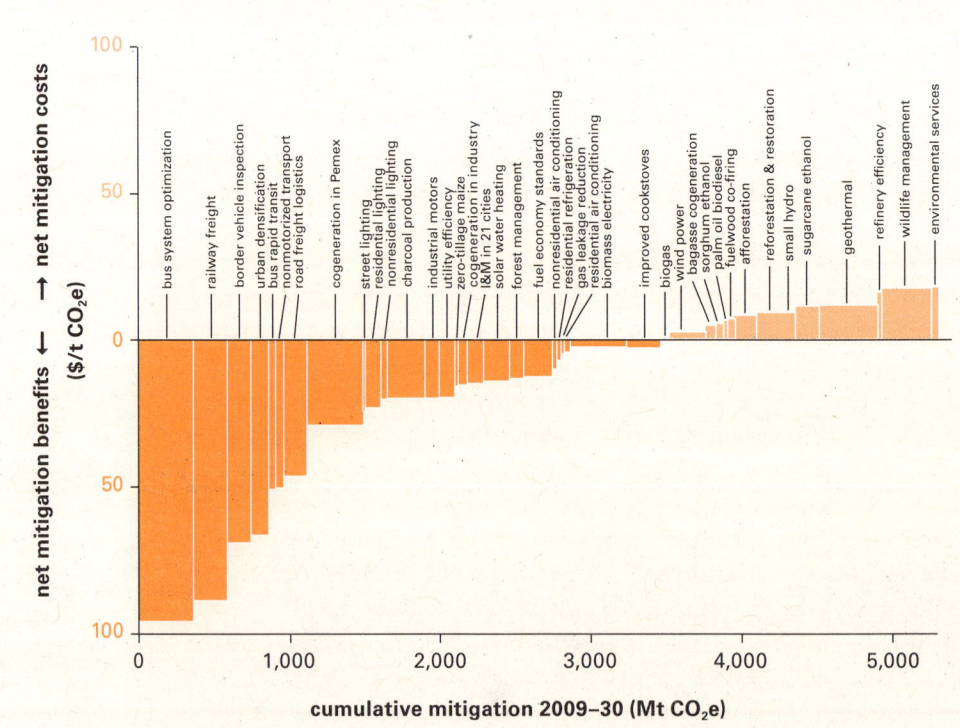

Source: Authors, based on MEDEC study results.

Among the interventions with the greatest total emissions abatement potential are geothermal electricity (393 Mt CO_2e), cogeneration in Pemex (387 Mt), biomass electricity (376 Mt), bus system optimization (360 Mt), wind power (240 Mt), improved cookstoves (222 Mt), and higher fuel economy standards (195 Mt). Together these seven interventions account for about 40 percent of the overall emissions-reduction potential of all MEDEC interventions.

The interventions with the highest benefit per ton of CO_2e abated are on the lefthand side of the marginal abatement cost curve. They include bus system optimization, road and railway freight logistics optimization, fuel economy standards, border vehicle inspection, urban densification, improved residential lighting, cogeneration in Pemex, and electric utility efficiency improvements.

Twenty-six interventions have negative net costs (that is, net benefits); together they account for about 65 percent of the overall emissions reduction potential of the interventions analyzed. Thirty-five interventions (including the 26 no-regrets interventions) could be achieved at a cost at or below \$10/t CO_2e. Together they account for 82 percent of the total emissions reduction potential of MEDEC interventions.

Putting the reduction potential and net incremental cost criteria together allows a first-order prioritization of low-carbon interventions (figure 7.4). All other things equal, the objective of a low-carbon program would be to promote projects with high emissions reduction potential and a net economic benefit.

Figure 7.4   Criteria for Selecting Low-Carbon Interventions

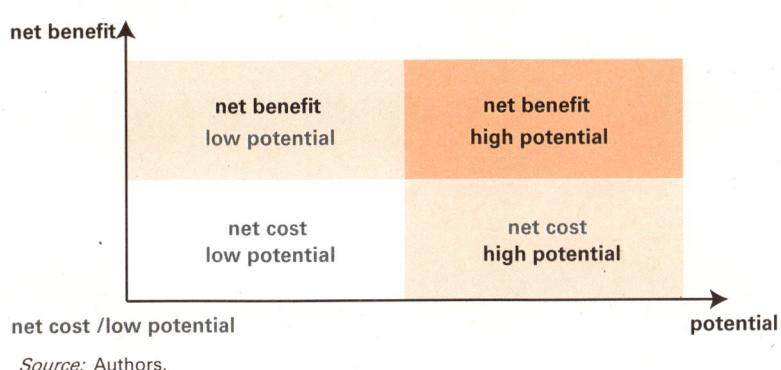

Source: Authors.

Macroeconomic Impact of MEDEC Interventions

The macroeconomic model developed by Boyd and Ibarrarán (2008) was used to assess the potential impacts of implementing MEDEC interventions on the Mexican economy. The outputs from MEDEC interventions (investment, operating and other costs, benefits) were scaled and integrated into a computable general equilibrium (CGE) model of the Mexican economy. The results from the MEDEC low-carbon scenario were compared in the CGE model with a baseline scenario using the same growth rate and other underlying variables. The CGE analysis allows an assessment of the impact of MEDEC low-carbon interventions on economic growth, the distribution of income, the level of economic welfare, the level of government revenue, the balance of trade, and the size of investment and capital in Mexico between 2008 and 2030.

The overall economic impact of implementing the MEDEC interventions was found to increase the overall level of GDP by as much as 5 percent in 2030. Under the MEDEC scenario, the level of overall investment in the economy climbs considerably, as does the final level of the capital stock. In the model, the respective investments by the government and the private sector are calculated according to the MEDEC interventions; the production functions in the model are revised over time to reflect the general increase in the efficiency of energy use. Government revenue rises slightly in the MEDEC scenario, indicating that the negative effect of subsidizing various low-carbon programs is more than compensated for by the increased aggregate tax revenues generated by an increase in the level of GDP. The overall increase in GDP is by no means evenly distributed: the agricultural and forestry sectors are by far the biggest winners. The impact on the level of welfare is progressive: per capita income grows for all income groups, with the greatest increase accruing to the lowest deciles (table 7.3).

Table 7.3 Combined Effect of MEDEC Interventions on the Mexican Economy
Percentage change with respect to the baseline

Parameter	2020	2030
GDP	2.06	5.58
Total investment under MEDEC scenario	7.04	15.82
Government spending under MEDEC scenario	−0.70	1.35
Final capital stock in the economy	—	7.55
Cumulative welfare		
Deciles 1–2	—	3.19
Deciles 3–5	—	2.96
Deciles 6–8	—	1.87
Deciles 9–10	—	0.84

Source: Authors.

Note: — = not available.

Notes

1. LEAP is a Windows-based software system designed for bottom-up energy and environmental policy analysis. It was developed and supported by the Stockholm Environment Institute U.S. Center (see www.energycommunity.org/).

2. The Mexican government changed its planning prospects to reflect the current financial crisis. It is considering a lower GDP growth rate to 2017. Given the long-term nature of the MEDEC modeling exercise, the study continues to assume the same long-run average GDP growth rate.

3. In 2007 the GDP carbon intensities of the United States and Japan were 0.53 kg CO_2e and 0.30 kg CO_2e per dollar, respectively; per capita CO_2e emissions were 24 t CO_2 and 11 t CO_2, respectively.

4. The magnitude of emissions reduction under the MEDEC low-carbon scenario is not highly dependent on the baseline assumption of a substantial increase in coal consumption. If natural gas were the primary incremental fuel for power generation under the baseline, the majority of MEDEC low-carbon interventions in the electricity sector would substitute for natural gas, thus slightly reducing the emissions reduction potential relative to coal. Given the expectation that natural gas generation (much of it imported as liquefied natural gas) would be more expensive than coal, the incremental cost for alternative low-carbon interventions for gas would be even lower, which would promote substitution. If the baseline were less coal intensive, the overall level of CO_2e emissions would decline; overall emissions in the baseline could be lower in 2030 than in 2008, but the MEDEC scenario emissions would be essentially the same.

5. The sectors correspond to the chapters of this report; they do not indicate where the emissions occur. For example, a number of interventions in the energy end-use sectors reduce emissions in the electricity sector.

6. Based on the "Methane to Markets" program in which Mexico is participating, methane leakage in the natural gas transmission and distribution system may be considerably underestimated. If this is the case, oil and gas sector emissions in the baseline scenario—and the potential for reduction—may be much larger.

7. Among the high-priority interventions not evaluated by MEDEC are those in waste management, such as landfill gas collection and urban recycling programs.

Elements of a Low-Carbon Development Program

There appears to be significant potential for Mexico to reduce its greenhouse gas emissions at fairly low cost. Based on the analysis, Mexico could keep its emissions relatively constant over the coming two decades while maintaining steady economic growth by following a low-carbon development pathway.

Although the MEDEC scenario assumes an aggressive program of low-carbon policies and investments, the magnitude of the emissions reductions obtained would appear to understate actual reductions, because of several conservative assumptions: only 40 of the many possible interventions are considered; the baseline assumes a rapid increase in fossil energy use by the transport and power sectors; and no major improvements in technology or reductions in technology costs are assumed. Moreover, nearly two-thirds of the interventions included involve actual cost savings relative to the baseline case, excluding externalities or transaction costs.

High-Priority Areas

Which sectors hold the most promise for reducing emissions at a low cost? High-priority areas for greenhouse gas reduction include interventions in the transport, electric power, forestry, and energy-efficiency sectors.

Transport

A substantial proportion of emissions reduction potential lies in the road transport subsector, the largest and fastest-growing emission sector in Mexico. Increasing the modal share of public and collective (as well as nonmotorized) transport in urban areas and raising the overall fuel efficiency of the vehicle fleet (for both passengers and freight) will be critical to reducing future road transport emissions.

Electricity

Given that Mexico will likely more than double its total power-generating capacity by 2030, it is important that new capacity be as efficient and low-carbon as possible. Based on international costs, it is possible that at least half of the new installed power capacity could be coal fired under the baseline. Mexico has significant cogeneration potential in industry (including in the oil and gas sector) and renewable energy resources (especially wind power in Oaxaca) that could begin to supply large amounts of power within the next five years at costs lower than Mexico's current marginal costs of electricity. Over the medium (5–10 years) to longer (more than 10 years) term, Mexico could develop significant renewable energy resources (hydro, wind, geothermal, solar, biomass), in many cases at low cost, that could be part of a low-carbon power development strategy.

Forestry

Although energy-related emissions dominate Mexico's current and projected CO_2e trajectories, the forestry sector provides the single greatest potential for reducing greenhouse gas emissions over the coming decades. Forestry interventions are generally more costly than those in transport or energy efficiency (on a \$/t CO_2e reduced basis), but most interventions that combine the reduction of deforestation and forest degradation benefit with the productive use of biomass, especially for energy purposes, have net benefits.

Energy End-Use

This study confirms the conclusions of other analyses that show that the overall potential for low-cost mitigation in the energy end-use sectors in Mexico is high in all sectors. The measures assessed for the study had the highest financial rates of return of any sector, as well as high economic rates of return without considering health benefits or other co-benefits, such as energy security or increased competitiveness.

"Feasibility" and Barriers to Implementation

What does it mean for a low-carbon intervention to be feasible? Almost all of the MEDEC interventions included in the low-carbon scenario have already been implemented in Mexico as regular investment projects or pilot programs, thus demonstrating their feasibility, at least on a limited scale. For many of the interventions, it is precisely the scale-up from an individual project to a broader program that is needed; scaling up such interventions typically involves changes in policies, institutions, and behaviors. Regulatory policy and incentives for investment in energy efficiency and renewable energy may not exist for implementing a low-carbon intervention on a wider scale. For instance, CFE and the private sector have implemented a number of wind projects in Oaxaca (many for self-supply), but the general policies needed to promote private sector provision of wind power to the grid are not yet mature in Mexico.[1]

Just because an intervention has positive net economic benefits and is feasible does not mean that it will happen automatically. Positive net economic benefits imply that the overall benefits for society of the project are greater than the costs; it says little about who the winners and potential losers are or whether the project has the political support to be approved and implemented. As highlighted in the sectoral analyses, a number of barriers—ranging from inexperience and the lack of information among suppliers and consumers to incompatibility with industry norms or government regulations—inhibit low-carbon interventions from being undertaken on a large scale. Many of the interventions evaluated in this study face a variety of market and nonmarket barriers, such as the high transaction costs associated with small projects or principal-agent problems, in which the beneficiary and the investor have different interests.

Two of the greatest challenges that Mexico and other countries will face in implementing a larger number of low-carbon interventions of the type evaluated in this study are financing the often larger upfront costs of low-carbon interventions and putting in place supportive policies and programs. Although a majority of the interventions have positive net present values, many low-carbon projects will require larger upfront investment in plant and equipment. Policies to promote low-carbon interventions exist, but new policies or changes in existing ones will be needed to accelerate the implementation of such interventions.

For both public and private decision making, upfront investment costs can be a major impediment to implementation. In many cases, low-carbon interventions, such as energy efficiency and renewable energy projects, have higher initial investment costs that are compensated by lower fuel and operating costs. But even if the life-cycle costs are lower, higher upfront investment costs often inhibit such investments from being approved and implemented, especially where credit markets are not well developed or implicit discount rates are high (that is, credit is expensive). Therefore, in addition to financial and economic analysis, it is important to assess the investment requirements for low-carbon interventions and to identify potential investment financing sources.

The marginal abatement cost curve shown in figure 7.3 does not indicate the level of investment needed for each investment. Those costs are presented in figure 8.1. The interventions are presented in the same rank order as in the marginal abatement cost curve, ranging from lowest net cost (highest net benefit) to highest net cost. The width of each bar measures the total reduction potential; the area of each bar represents the total investment for that intervention.

These results tell a very different story from that told by the marginal abatement cost curve. Some interventions with large emissions reductions and low net costs have very large investment requirements; other interventions have low or minimal investment requirements. Not surprisingly, the largest investment requirements are for large-scale and capital-intensive interventions, such as renewable energy projects (geothermal, wind, bio-

Figure 8.1 Marginal Abatement Investment Curve

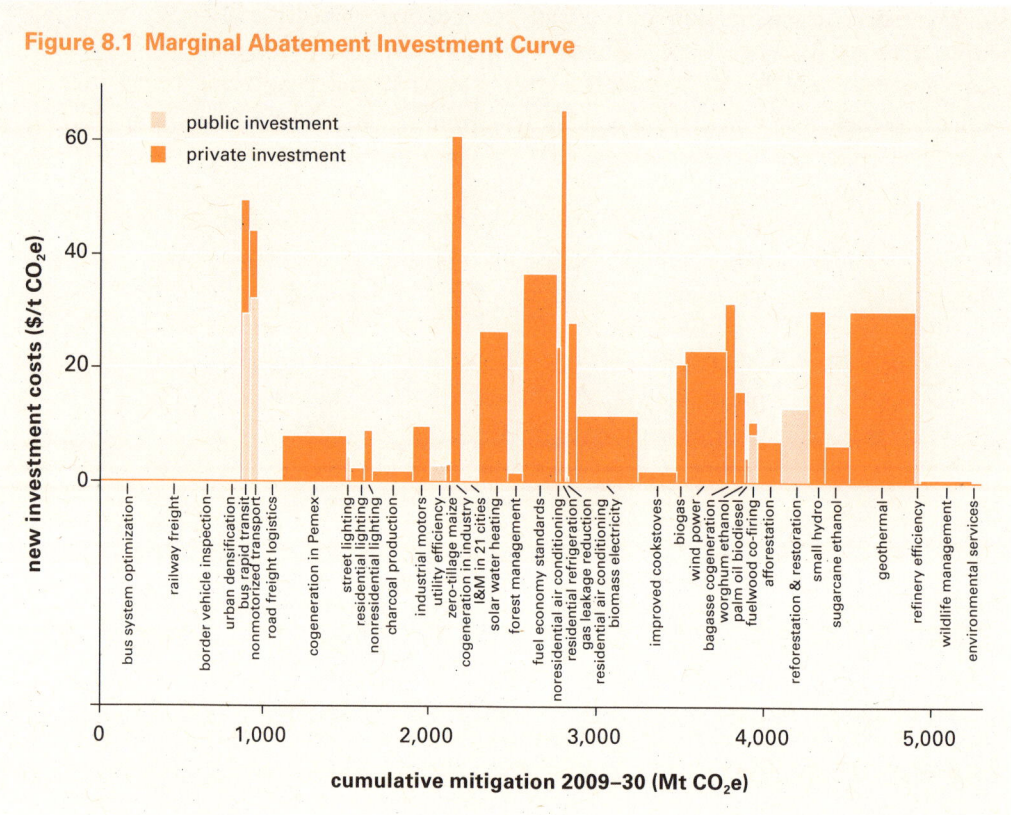

Source: Authors.

mass electricity, small hydro, and solar water heating); energy efficiency (cogeneration, refinery efficiency, and residential refrigeration); and transport (bus rapid transit and fuel economy standards).

But not all low-carbon interventions have higher investment costs. Among the interventions that may not have high investment costs are those related to improvements in operational or organizational efficiency (bus system optimization, road freight logistics); better utilization of existing infrastructure (railway freight);[2] or adoption of vehicle inspection programs.

In some cases, the barrier is not direct financial costs or investment hurdles but rather the costs of developing, passing, and enforcing new regulations, such as efficiency standards or operational norms for new and existing equipment. Even though all suppliers may be subject to the new standards, manufacturers may oppose them out of fear that they will drive up production costs, reducing sales. Information about the benefits of the program—for both producers and consumers—could help overcome opposition.

Inspection programs for in-use vehicles are interventions that would also have low investment costs. Such programs can help keep highly polluting and out-of-tune vehicles off the road, reducing both local pollutants and CO_2e. The costs of the program can be covered by nominal fees for vehicle owners through regular inspections.

Implementing all of the MEDEC interventions over the period 2009–30 would require investment of $64.5 billion, or about $3 billion a year (table 8.1). This level of investment represents about 0.4 percent of Mexico's current GDP.

Table 8.1 MEDEC Investment Requirements to 2030
$ millions

Sector	New investment	Forgone investment	Net investment
Electric power	21,406	10,933	10,473
Oil and gas	4,637	1,482	3,155
Energy end-use	15,771	9,898	5,873
Transport	11,729	36,249	−24,520[a]
Agriculture and forestry	10,928	3,699	7,229
Total	64,471	62,261	2,210

Source: Authors.
a. A negative net investment means that new investments under the low-carbon scenario are less than the avoided (forgone) investment under the baseline scenario.

In addition to the new investments required for MEDEC interventions, some investments made under the baseline would be forgone. Investments in low-carbon electric power capacity and energy efficiency (lighting, air conditioning, refrigeration), for example, would replace investments in power plants fired by natural gas or coal. Investment in buses and infrastructure would be reduced (as a result of bus system optimization); a large number of smaller buses would be replaced by large articulated buses in BRTs; and the need for trucks and freight infrastructure would fall (as a result of the optimization of road freight logistics). Overall, the value of the avoided transport investments associated with the MEDEC interventions is estimated to be worth more than three times the value of new investments, resulting in overall negative net investment under the MEDEC transport scenario. (Because many of the avoided investments accrue to different actors, it is not meaningful from a project perspective to subtract avoided investments from new investments. However, the "net" investment numbers presented in table 8.1 do reflect the investment requirements for Mexico as a whole.)

Financing Low-Carbon Interventions

Investment in low-carbon development need not come from the government (figure 8.1). Even under current budgeting practices, the vast majority of the interventions—including most of the energy-efficiency investments—would be financed by the private sector and households (table 8.2).

Table 8.2 Low-Carbon Interventions by Financing Source

Private sector	Household	Public sector[a]
Commercial energy efficiency	Residential energy efficiency	Street lighting
Industrial energy efficiency	Solar water heating	Public services efficiency
Cogeneration in industry, including sugar factories	Zero-tillage maize	Reforestation and restoration
IPPs for renewables (wind, biomass)	New vehicles	Transport infrastructure
	Vehicle I&M	Geothermal power
Buses		Oil and gas investments
Liquid biofuels		

Source: Authors.
Note: I&M = inspection and maintenance.
a. Worldwide, many public sector investments are financed through concession schemes with private contractors or operators, including for power generation, oil and gas, public transportation, and other public utilities (water and sanitation).

Government support is important for many public infrastructure investments, and government subsidy programs can and should be designed to introduce and accelerate the adoption of some low-carbon interventions. It is also possible to shift more investment for traditionally public services and infrastructure, such as urban transportation or the energy sector, to the private sector through public concessions or other types of public-private partnerships. Among the specific areas in which the private sector could become more active with changes in regulatory policies are public sector energy efficiency and renewable energy production. In addition, improving the efficiency of public financing—especially in the petroleum and electricity sectors in Mexico—can reduce the cost and risk to the government.

Mexico is unique among middle-income countries in that the energy industry—including at the retail distribution level—is largely in the hands of three large state-owned companies: Pemex, CFE, and LyFC. The role of public and private sector investment will be particularly important in Mexico in the energy sector given the dominance of state-owned companies and the limitations (including in the constitution) on private sector investment. Mexico can provide a conducive environment for investment in the energy sector without being the primary investor itself. As some state-owned oil and power companies in other countries have shown, doing so does not necessarily mean sacrificing national sovereignty over the ownership of strategic natural resources. What is needed to attract investment in the electric power and the oil and gas sectors are stable environments with clear rules that allow contracting according to international best practices. There is substantial room for improving the operational and investment efficiency of the state-owned energy industry in Mexico; climate change concerns can provide additional leverage for making such improvements.

Despite recent turmoil in international financial markets, Mexico will remain an attractive country for private sector investment in the energy field, and there will likely be increasing attention worldwide to opportunities in renewable energy, energy efficiency, and sustainable transport. There is thus considerable room to involve the private sector in these sectors in Mexico. Recent government reforms aimed at improving the efficiency of the state energy sector represent a positive step. The dramatic increase in the number of independent power producers in Mexico since the mid-1990s demonstrates the potential for involving the private sector (even if the model chosen involved higher risk and higher cost for the public sector than is typical worldwide).

Another area in which government investment is important is research and development on low-carbon technologies and interventions. In many areas, Mexico can take advantage of the technological advances made in other countries that will help lower greenhouse gas emissions, including technologies currently on the horizon, such as carbon capture and storage, and technologies that are yet to be developed. In other areas, such as large-scale wind machines, Mexico has a comparative advantage; the government should help promote research and industrial development in such areas. Research areas that are more important to Mexico than to other countries—such as developing energy-efficient residential building standards in hot and dry climates—may also warrant dedicated effort.

Policies for Low-Carbon Development

Many of the high-priority MEDEC interventions will require changes in policies before they can be implemented on a large scale (box 8.1). Some policy barriers can be removed or reduced through specific new regulations directed at a class of interventions, such as renewable energy legislation; other barriers, such as the impact of low energy prices on energy-efficiency investments, are economywide. Some low-carbon interventions—such as those in urban transport—will require increased coordination among multiple government agents and across different levels of government. A number of interventions require both longer-range planning by the government and more continuity across the six-year administrative periods at the federal and state levels.

Many recommended policies—such as contracts with independent power producers or ESCOs—are not new in Mexico. They could be improved and extended to promote low-carbon development through energy efficiency and renewable energy (which would also have energy diversification and environmental protection benefits). Among the problems for renewable energy projects has been the low planning prices (including the exclusion of externalities) for fossil fuels, the lack of adequate capacity recognition for intermittent renewables, and the inability to adjust procurement procedures to the requirements of renewable energy projects. Preestablishing small power purchase agreements would promote electricity sales to the grid from small or intermittent renewable energy or cogeneration producers.

Box 8.1 Policies to Support Low-Carbon Development

A variety of policies could support low-carbon development in Mexico. Seven of them are described here.

- **Electric power from renewables.** Promotion policies, such as predefined contracts, and tariffs ("feed-in tariffs") that permit and actively encourage small generators to produce and sell electricity to the grid by reducing project development risks, would increase power, much of it at a lower cost than CFE currently pays. Establishing small power purchase agreements would be a useful first step.
- **Energy-efficiency standards.** The establishment or improvement of existing minimum efficiency standards for widely used equipment (motors, pumps, lighting, boilers, furnaces), appliances (air conditioners, refrigerators), and vehicles (cars, trucks, buses) would reduce per unit energy consumption. Standards need to be complemented by measures to ensure the efficiency of used vehicles and equipment, such as vehicle inspection and maintenance programs, and cash payments for scrappage of vehicles and appliances.[a]
- **Energy pricing.** In light of the regressive nature of energy subsidies in Mexico, reductions in implicit subsidies for middle- and high-income residential electricity consumers would have an immediate impact on reducing electricity consumption in Mexico while improving the distributive effect of energy pricing (see box 4.2). Raising gasoline prices, which have been stable or have fallen over the past 20 years, would have a direct effect on the use of private automobiles, a primary contributor to Mexico's increasing greenhouse gas emissions over the past 25 years.
- **Changes in public procurement rules.** Energy efficiency in many public facilities (schools, hospitals, government buildings, water supply and sanitation) is restricted by the inability of public agencies to sign contracts with private energy-efficiency companies for more than one year. Revision of public procurement rules would help public institutions save energy and reduce their operating costs.

(continued)

Recent legal changes have removed barriers to tap Pemex's cogeneration potential—which represents more than 6 percent of total installed capacity in Mexico—but there is still a need to establish a regulatory framework that enables these projects to offer energy and capacity to the grid at scale and with adequate incentives.

The Importance of Co-Benefits

Positive externalities (co-benefits) can be large for certain types of mitigation measures; their inclusion can help justify low-carbon interventions. The fact that positive externalities are not included in the comparative economic analysis presented in chapter 7 means that projects that reduce fossil fuel consumption or protect forests would have even higher net economic returns if health or ecological benefits were included.[3]

Transport interventions that reduce overall transport intensity or improve vehicle efficiency can have significant positive impacts on acute

Box 8.1 Policies to Support Low-Carbon Development *(continued)*

- **Urban planning and public transport.** Complementary regulations and coordinated actions by federal, state, and municipal government agencies are needed to promote urban planning that reduces overall transport demands (high-density zoning, radial corridors) and provides convenient, accessible, and safe public transport infrastructure, including areas for pedestrians and bicycles.

- **Forestry programs.** Policies to manage and protect native forests—such as those that control illegal logging, prevent fires, and manage pests—will yield local and global environmental benefits. Other measures to reduce deforestation and promote reforestation/afforestation programs include community forestry programs.

- **Air quality standards.** Improved fuel quality standards and better enforcement of air quality standards could provide cost-effective CO_2 reduction. Improved fuel quality—principally for gasoline, diesel, and fuel oil—would help meet Mexico's ambient air quality standards; by allowing better engine performance, it could reduce CO_2 emissions. Inspection and maintenance programs help keep grossly out of tune vehicles off the road, improving local air quality and raising fuel efficiency. Alternatively, air quality standards could be defined as atmospheric pollutant concentration standards rather than vehicle emission standards, and local authorities be made responsible for meeting them; this would enable the authorities to seek—according to the local context—other means of compliance beside vehicle emissions, such as public transportation or nonmotorized transport, which could also have important greenhouse gas mitigation effects. All measures would help air pollution "nonattainment" areas in Mexico meet air quality standards.

---

a. In the case of refrigerators, international experience shows that when people purchase a new refrigerator and the old one is not taken in trade or removed, it can end up as a second refrigerator in the same household or transferred to another household, thus resulting in an overall increase in electricity use for refrigeration and negating the potential efficiency gains of the new appliance.

respiratory disease and asthma. Vehicle inspection and licensing programs at the national level would result in large energy savings for vehicle owners and help meet Mexico's air quality standards. This is an area of particular importance for Mexico given the large numbers of used vehicles that enter the country from the United States each year.[4] Forestry projects—including avoided deforestation and reforestation—can generate large environmental benefits in terms of soil conservation, water quality, and ecosystem preservation (externalities that were not estimated in MEDEC), in addition to providing employment and income for rural communities.

Air pollution in Mexico City is an example of a negative externality that has attracted significant public attention and resulted in a substantial political response. During the 1990s, Mexico City implemented many measures to reduce air pollution. These measures reduced the number of days each year that air quality standards are violated. The climate change mitigation interventions outlined in this report provide air pollution reduction as a co-benefit (the primary objective is to reduce overall energy consumption and greenhouse gas emissions). Many projects currently being promoted as "climate change" projects had previously been advocated for their energy security (renewables and energy efficiency) or local health and environment (reforestation and urban transportation) benefits.

As elsewhere in the world, co-benefits in Mexico are typically not included in cost-benefit analysis or are undervalued in public decision making. Internalizing such benefits and costs—through pollution charges for air pollution or payments for environmental services, for example—is likely to lead to more efficient outcomes.

Near-Term Actions

As the government of Mexico moves forward with its climate change mitigation program, it is important that it prioritize near-term interventions. This study recommends that priority be given to interventions with the following characteristics:

- Significant emissions reduction potential
- Positive economic rates of return, including large co-benefits
- Successful demonstration at commercial scale in Mexico or internationally
- Low investment costs and the ability to obtain financing.

An additional consideration, in light of the international financial crisis of 2008–09, is that low-carbon interventions should have positive employment and secondary development effects. Initial evidence suggests that investments that contribute to improving the capital stock have the greatest impact on employment (additional research on this topic is warranted).[5]

The MEDEC interventions were limited to existing commercial technologies; all are therefore available today. All of the energy-efficiency interventions, plus those involving efficiency improvements in the power sector (cogeneration, utility efficiency) are technically ready, and all have substantial commercial demonstrations in Mexico. Some technologies—such as biomass power generation—have been demonstrated at scale abroad but not in Mexico; these interventions may need several years of market development to ramp up. The benefits of interventions involving changes in urban infrastructure—roads, buildings, housing, pedestrian facilities—will take time to reap, but all could be started immediately. Because the majority of interventions evaluated cost less than $10/t CO_2e (and no interventions were considered that cost more than $25/t CO_2e), most are economically viable today or would be so in the near future, assuming the development of a widespread international carbon market in which Mexico can participate.[6]

A final important criterion for implementation in the near term is that the legal, regulatory, and institutional barriers to implementation be surmountable. The litmus test for MEDEC interventions has been that they have already been successfully undertaken in Mexico or abroad. Most of the MEDEC interventions meet these criteria. Institutional barriers, such as those discussed with respect to the energy sector, remain and will continue to inhibit investments and efficiency improvements, but all could be overcome with modest changes in regulations governing the energy industry.

Several low-carbon interventions that meet the criteria of potential, cost, and feasibility could be implemented in the short to medium term (one to five years). Some of these interventions, such as BRT, are already being scaled up. Based on projects in Mexico City and pioneered in other parts of Latin America, BRT is being expanded to other routes in Mexico City as well as in other large cities in Mexico. Other examples of projects that could be scaled up in the near term include residential lighting programs developed under FIDE, wind farms in Oaxaca based on CFE's pilots, forest management based on the Los Tuxtlas project in Veracruz, cogeneration in Pemex refineries based on the project at the Nuevo Pemex Refinery, and fuel economy standards for new vehicles and inspection programs for used vehicles (table 8.3).

International Support

Several international mechanisms could support Mexico's low-carbon development program. An international agreement to set emissions limits on industrial countries and to extend the carbon market mechanisms is a necessary ingredient to maintaining the sale of carbon credits by developing countries. International political momentum for adopting climate change mitigation actions has been growing over the past several years, and the stage has been set for a new agreement that will further motivate actions to reduce greenhouse gas reduction by both industrial and developing countries, with developing countries benefiting from a carbon trading system.

Based on the experience gained through clean development mechanism projects—both positive and negative—it is likely that the private carbon market will continue to focus on projects that are relatively easy to finance. These will include projects to reduce methane, such as landfill gas and animal waste projects. They will also likely include small and discrete projects whose emission reductions are relatively easy to verify and monitor (such as single-technology interventions in the energy sector). Revision of the rules governing the clean development mechanism or a subsequent replacement mechanism to allow more flexibility for promoting mitigation projects in developing countries, including a move toward a policy and programmatic approach, is needed.

Programs to support mitigation of climate change supported by bilateral and international organizations, including those governed by the UNFCCC, will seek to expand the current mitigation agenda to project areas that have not been the mainstay of the private carbon market. There is a need, for example, to expand the coverage of carbon markets to include more land-use projects, an important source of emissions reductions. Such projects have relatively modest financial costs and could benefit from the political support that carbon revenues could provide.

Table 8.3 Potential Near-Term Interventions

Intervention	Total new investment ($ millions)	Total emissions reduction (Mt CO_2e)	Maximum annual emissions reduction (Mt CO_2e)	Mitigation cost or benefit ($/t CO_2e)	Implementation time frame
Utility efficiency	286	103	6	19 (benefit)	Short term
Wind power	5,549	240	23	3 (cost)	Short/ medium term
Cogeneration in Pemex	3,068	387	27	29 (benefit)	Short/ medium term
Residential lighting	237	100	6	23 (benefit)	Short term
Solar water heating	4,464	169	19	14 (benefit)	Short/ medium term
Nonresidential lighting	420	47	5	20 (benefit)	Short term
Improved cookstoves	434	222	19	2 (benefit)	Short term
Border vehicle inspection	0	166	11	69 (benefit)	Short term
Bus rapid transit	2,332	47	4	51 (benefit)	Short term
I&M in 21 cities	0	109	11	15 (benefit)	Short term
Forest management	148	92	8	13 (benefit)	Short term
Bus system optimization	0	360	32	97 (benefit)	Short/ medium term
Nonmotorized transport	2,252	51	6	50 (benefit)	Short/ medium term
Road freight logistics	0	157	14	46 (benefit)	Short/ medium term
Fuel economy standards	7,145	195	20	12 (benefit)	Short/ medium term
Afforestation	1,084	153	14	8 (cost)	Short/ medium term
Reforestation & restoration	2,229	169	22	9 (cost)	Short/ medium term
Total	29,648	2,767	247		

Source: Authors.
Note: I&M = inspection and maintenance.

Another area that has not been sufficiently supported by the carbon market, and that is a high priority for Mexico and other middle-income countries, is road transport. The Global Environment Facility and new initiatives such as the Clean Investment Funds are increasingly interested in greenhouse gas mitigation in such areas as sustainable transport programs and other programs that have not seen much involvement by either the private carbon market or public climate change mitigation programs. MEDEC provides additional evidence of high-priority interventions in the transport sector. Based on findings from this study, Mexico has submitted a proposal to tap funding from the Clean Investment Funds for sustainable transport, energy efficiency, and renewable energy.

Notes

1. In some cases, initial pilot projects were funded in part with grant resources, such as funds from the Global Environment Facility.

2. The investment costs for this intervention were assumed to be zero (based on better utilization of existing rail infrastructure). In reality, a substantial increase in railway freight would involve investment costs in engines, cars, and probably track.

3. For examples of the health benefits from transport and improved cookstove interventions, see boxes 4.1 and 6.1.

4. The flow of older and dirtier vehicles into Mexican states with lax enforcement of environmental standards and weak vehicle inspection and maintenance programs is probably an example in which free trade and differential environmental standards can worsen environmental quality in the receiving country.

5. The macroeconomic modeling using the computable general equilibrium model of Mexico yields evidence of a positive correlation between the low-carbon scenario and employment. Unemployment was lowest in the scenarios that resulted in the largest increase in new capital stock.

6. More than four-fifths of the emissions reduction potential of MEDEC interventions had a cost of less than $10/t CO_2e, without considering positive externalities.

Summary of MEDEC Interventions

Table A1 Estimated Investment, Emissions Reduction, and Net Abatement Cost of MEDEC Interventions

Intervention	Sector	New investment ($ millions)	Total emissions reduction (Mt CO_2e)	Maximum annual emissions reduction (Mt CO_2e)	Net cost or benefit of mitigation ($/t CO_2e)
Bus system optimization	Transport	*	−360	−31.5	−97
Railway freight	Transport	0	−220	−19.2	−89
Border vehicle inspection	Transport	0	−166	−11.2	−69
Urban densification	Transport	*	−117	−14.3	−66
Bus rapid transit	Transport	2,333	−47	−4.2	−51
Nonmotorized transport	Transport	2,252	−51	−5.8	−50
Road freight logistics	Transport	0	−157	−13.8	−46
Cogeneration in Pemex	Oil and gas	3,068	−387	−26.7	−29
Street lighting	Energy efficiency	39	−9	−0.9	−24
Residential lighting	Energy efficiency	237	−100	−5.7	−23
Nonresidential lighting	Energy efficiency	420	−47	−4.7	−20
Charcoal production	A&F	416	−248	−22.6	−20
Utility efficiency	Electricity	286	−103	−6.2	−19
Industrial motors	Energy efficiency	907	−94	−6.0	−19
Cogeneration in industry	Energy efficiency	3,738	−61	−6.5	−15
Zero-tillage maize	A&F	74	−25	−2.2	−15
Solar water heating	Energy efficiency	4,464	−169	−18.9	−14
I&M in 21 cities	Transport	0	−109	−10.6	−14

(continued)

Table A1 Estimated Investment, Emissions Reduction, and Net Abatement Cost of MEDEC Interventions *(continued)*

Intervention	Sector	New investment ($ millions)	Total emissions reduction (Mt CO₂e)	Maximum annual emissions reduction (Mt CO₂e)	Net cost or benefit of mitigation ($/t CO₂e)
Forest management	A&F	148	−92	−7.8	−13
Fuel economy standards	Transport	7,145	−195	−20.1	−12
Nonresidential air conditioning	Energy efficiency	589	−25	−1.7	−10
Residential refrigeration	Energy efficiency	1,907	−29	−3.3	−7
Residential air conditioning	Energy efficiency	1,174	−42	−2.6	−4
Gas leakage reduction	Oil and gas	16	−17	−0.8	−4
Biomass electricity	A&F	4,254	−376	−35.1	−2
Improved cookstoves	Energy efficiency	434	−222	−19.4	−2
Biogas	Electricity	1,141	−55	−5.4	1
Windpower	Electricity	5,549	−240	−23.0	3
Bagasse cogeneration	Energy efficiency	1,860	−59	−6.0	5
Sorghum ethanol	A&F	991	−62	−5.1	5
Palm oil biodiesel	A&F	99	−24	−2.4	6
Fuelwood co-firing	A&F	454	−43	−2.4	7
Afforestation	A&F	1,084	−153	−13.8	8
Small hydropower	Electricity	2,634	−86	−8.8	9
Reforestation and restoration	A&F	2,229	−169	−22.4	9
Sugarcane ethanol	A&F	1,011	−150	−16.8	11
Geothermal power	Electricity	11,797	−393	−48.0	12
Refinery efficiency	Oil and gas	1,553	−31	−2.5	17
Wildlife management	A&F	169	−316	−27.0	18
Environmental services	A&F	0	−51	−4.4	18

Source: Authors.
Note: A&F = agriculture and forestry; I&M = inspection and maintenance.
* New investment for intervention was negative, that is, less than under the baseline.

Summary of Benefit-Cost Analysis Methodology

Cost-effectiveness is defined as the present value (in 2008) of the net benefit of reducing (avoiding) 1 ton of CO_2–equivalent emissions (\$/t CO_2e) by implementing a particular option (definitions of CO_2e are from IPCC 2007). For each intervention, the annual emissions reductions (in t CO_2e) are summed to calculate total emissions reduction, and the stream of annual net cost is discounted at 10 percent a year to arrive at the present value of the net cost. The cost-effectiveness ratio is then calculated by dividing the second amount by the first. Using the cost-effectiveness form of benefit-cost analysis allows the analyst to avoid directly estimating the marginal value (damage function) for each additional ton of CO_2e added to the atmosphere. At the same time, the cost per ton of CO_2e estimate for each option considered provides a convenient comparison against which estimates of CO_2e damage functions or carbon market prices can be made.

The net benefit of a mitigation option is calculated by subtracting the direct financial costs from the direct benefits of implementing it. Examples of direct benefits include energy cost savings or travel time and cost savings. Indirect benefits, such as environmental externalities, are not quantified. The financial costs reflect economic opportunity costs to the extent that corrections were made for taxes and subsidies and that traded goods were assessed at their import and export parity values.

Pairwise comparisons are made between particular options and the baseline scenario (the alternative that presumably would have been pursued in the absence of the MEDEC program). Incremental net costs and incremental net greenhouse gas emissions are calculated by subtracting the costs (or greenhouse gas emissions) of the option from the costs (or greenhouse gas emissions) of the baseline case.

The analysis used a cost-effectiveness format in which "output" is counted but not valued in currency terms and "input" costs are measured and valued in constant (2008) U.S. dollars. The output in this case is tons of CO_2e avoided by the option (relative to emissions under the baseline alternative). Benefits are net of indirect co-benefits (see below). The cost per ton of net CO_2e emissions avoided (mitigated) by each option was then calculated.

In the cash-flow format, the annual emission of CO_2e appears as the annual flow of CO_2e in that year, but it adds to a "stock" of greenhouse gas in the atmosphere that will continue to be there at the end of the plan period (2030). As discounting and compounding are mathematical methods for converting a flow resource into a stock equivalent at the beginning or the end of the plan period, it would not be appropriate to compound or discount a number (tons of CO_2e) that already represents a stock value. Thus, the cost-effectiveness ratio that is calculated represents the cost per ton of CO_2e stock avoided or mitigated for the entire time it would have remained in the atmosphere.

Each option analyzed had a project life based on the economic (rather than the physical) life of the most important asset. Less important assets having longer lives than the project life have their remaining salvage value added back to the cash flow at the end of the project life. For assets that do not last as long as the most important asset and thus must be replaced from time to time during the project life, their investment value enters into the cash flow at more than one point. If the project life was not evenly divisible by the life of the shorter-lived asset, a salvage value related to the final replacement of the asset was added back at the end of the regular project life.

A series of similar projects makes up a *program*. Program duration is always 2009 to 2030, with projects usually starting at different dates. Most projects go beyond the end of the program (2030), either because they start after 2009 or because their assets (such as power plants) have economic lives exceeding the 22-year plan period. In this case, assets with remaining life after 2030 had their residual value added back to the cash flow for 2031. Residual value includes the net sale value of any assets plus the recapture of working capital stocks remaining when production is shut down, whether prematurely or at the exhaustion of the most important asset. Net sale value of remaining assets is called *salvage value* if the asset has unused life remaining. If the asset has come to the end of its useful economic life, the term *scrap value* is more commonly applied. The common convention with scrap values is to assume that removal costs are equal to market value of the scrap, suggesting a residual value equal to zero for those assets. Salvage values were not applied to assets being replaced under the low carbon program before they were fully depreciated, because doing so would imply continued use of the asset and continued emissions of greenhouse gases.

Greenhouse gas reduction options analyzed for the MEDEC portfolio were limited to technologies already in use or those realistically expected to come into use within five years. Moreover, no technological progress was presumed once the investment in the option went from "putty" to "clay."

A full-blown economic analysis normally starts with the financial cash flow of the most important stakeholder, to which it makes the following adjustments:

1. Delete direct transfer payments (taxes and subsidies).
2. Divide inputs and outputs between traded and nontraded goods (and services), and value the traded items at import and export parity equivalent values.
3. Use input-output analysis or other methods to trace and remove the indirect taxes and subsidies involved in supplying nontraded inputs to the project.
4. Convert nontraded outputs to willingness-to-pay values (which involves extensive analysis of the degree of market development and market distortions in some cases).
5. Determine quantitative measures of environmental spillovers, develop damage functions related to those spillovers, and determine willing-ness-to-pay values or willingness-to-accept compensation values for these externalities.

The MEDEC economic analysis involved only the first two of the five steps in this sequence. Because of the importance of environmental co-benefits in their sectors, the transport group and the electricity group attempted to complete step five as a side calculation, without including these co-benefits in calculating cost per ton of greenhouse gas mitigated.

The objective function for the MEDEC study is cost per ton of greenhouse gas reduction or mitigation. Non–CO_2 greenhouse gases are converted to CO_2 equivalents (CO_2e); other impacts are either converted to net costs or are ignored. Outputs that are produced in conjunction with greenhouse gas reductions are divided into direct and indirect co-benefit categories. Direct co-benefits (such as time savings and automobile expenditures saved by riders of urban transport or energy savings by users of energy-efficient household appliances) are included in the net cost calculation where fea-sible. Indirect co-benefits (such as environmental externalities) are counted where feasible, but their imputed values in willingness-to-pay terms are not included as co-benefits in the calculation of cost per ton of CO_2e reduction.

The cost-per-ton calculations do not include the additional organiza-tional and institutional interventions that might be required to overcome barriers to implementing an option. For example, the reduced-tillage option does not specify ownership of the equipment or organizational costs nec-essary to make the equipment available to the farmers who are expected to use these new practices; the costs do not include the information and education costs of encouraging the adoption of reduced tillage. The costs of household energy-efficiency options exclude the costs of organizing distribution; convincing households that the options are better than the high-carbon alternatives; and developing certification, service, and mainte-nance systems. These project costs cannot be calculated until interventions designed to remove existing barriers are identified. Omission of these costs plays a large part in explaining the negative net cost-per-ton outcomes for several of the options analyzed by MEDEC.

Positive net benefits (or negative net costs) of an investment option usually suggest the presence of barriers that prevent private parties or public agencies from acting in a way that cost-effectiveness calculations suggest makes economic sense. Without these barriers, profitable investments presumably would not be left on the table. The fact that no-regrets greenhouse gas reduction options exist suggests that that the remaining task is to identify the barriers that account for them, analyze the ability to surmount them, and design the requisite programs of interventions to remove, surmount, or skirt them. The surmountability of the barriers and the cost of interventions to surmount them then becomes the third criterion in rating the investment options (along with the net benefit per ton of greenhouse gas reduction provided by that option and the scope the option provides for reducing greenhouse gas).

Intervention Assumptions

This appendix describes the assumptions used throughout the analysis. It first introduces the general assumptions before introducing specific assumptions adopted in each sector.

The following general assumptions were used in the MEDEC analysis:

- MEDEC duration: 22 years (2009–30) (programs are made up of a series of projects, which may have different durations, usually according to the lifetime of their main assets)
- MEDEC year zero: 2008
- Discount rate for costs and externalities: 10 percent
- Discount rate for CO_2e emissions: 0
- Year of constant dollars: 2005
- GDP annual growth rate: 3.6 percent
- Average annual population growth: 0.6 percent
- Changes in technology: No major change in technology over the scenario period
- Net costs (or benefits): Sum of net present value of new public investment, new private investment, forgone investment, salvage value (salvage value in 2031 was calculated in a nonlinear way), energy costs (includes only fossil energy costs), other operations and maintenance costs, labor costs, and unpaid time costs (time savings were calculated using the minimum wage of $0.55/hour)
- Fuel prices: West Texas Intermediate crude oil price (about $53 per barrel in 2009; table C.1).
- Emission factors for fossil fuels: Standard IPCC factors for downstream (end of pipe) emissions. For upstream emissions, sources are Yan (2008) for LPG, gasoline, and diesel and Hondo (2005) for fuel oil, natural gas, and coal (coke upstream emissions are assumed to be equal to those for coal) (table C.2).

Table C.1 Fuel Cost Assumptions for MEDEC Interventions

Type of fuel	Cost in 2009 ($/GJ)	Annual cost increase (%)
Gasoline	15.98	0.567
Diesel	12.84	0.527
Fuel oil	7.39	0.403
LPG	12.09	0.469
Natural gas	7.85	0.190
Coal	2.07	0.471
Coke	15.02	0.471

Source: Authors.

Table C.2 Downstream and Upstream Emissions
t CO2e/GJ

Type of fuel	Downstream emissions	Upstream emissions
Gasoline	0.0693	0.0160
Diesel	0.0741	0.0173
Fuel oil	0.0774	0.0038
LPG	0.0631	0.0130
Natural gas	0.0561	0.0135
Coal	0.0946	0.0090
Coke	0.1082	0.0090

Source: Hondo 2005; IPCC 2007; Yan 2008.

Electricity Sector

According to the government's official outlook, electricity demand is expected to grow 4.9 percent a year through 2016 (SENER 2007). A growth rate of 3.9 percent a year was assumed for the period 2017–30.

The selection of power-generation technologies (additions and withdrawals) for the period 2007–16 is based on the official outlook. Technologies beyond 2016 are based on the following assumptions:

- Expansion is based on demand projections and meeting the load curve.
- Expansion is based on least-cost technology.
- Old power plants are withdrawn.
- Environmental requirements for criteria pollutants (particulates, SO_2, and NO_x) are met.

Investment costs are based on international values (World Bank 2008). Operations and maintenance costs and fuel consumption figures reflect Mexico's local conditions (CFE 2008a). Unit costs are the same regardless of scale (no economies of scale are considered). The cost of cooling water is assumed to be $0.679/m³. Table C.3 shows the costs assumed for the coal and natural gas technologies.

Table C.3 Baseline Technology Characteristics of Coal and Natural Gas

Characteristic	Coal, supercritical	Natural gas, combined cycle
Overnight private investment ($/[MWh/year])	321	203
Operations and maintenance ($/MWh)	6.490	4.080
Externalities ($/MWh)	1.859	0.580
Fuel use (GJ/MWh)	8.356	6.901
Capacity (MW/[MWh/year])	0.00015245	0.00014671

Source: World Bank 2008; CFE 2008a.
Note: Real investment occurs over several years; overnight investment is its equivalent in financial terms on the day the plant becomes operational.

These considerations led to a baseline scenario based primarily on coal, natural gas, and hydropower (table C.4). Under this scenario, coal would be used in power plants located in coastal areas near ports, providing base power; natural gas and hydropower would cover inland areas and intermediate and peak production. Given the availability of fossil fuels in Mexico, it is likely that most coal and gas in the baseline scenario would be imported.

Table C.4 Projected Energy Capacity and Generation under the Baseline Scenario

Source of energy	Capacity in 2008 (MW)	New capacity 2009–30 (MW)	Withdrawals 2009–30 (MW)	Capacity in 2030 (MW)	Generation in 2030 (GWh)
Natural gas	23,104	28,008	−4,095	47,016	293,353
Coal	4,718	26,391	0	31,108	208,783
Hydro	11,466	13,727	0	25,193	86,784
Fuel oil	12,830	0	−7,112	5,718	26,826
Geothermal	960	976	−150	1,785	13,890
Uranium	1,365	269	0	1,634	12,610
Natural gas cogeneration	2,069	314	0	2,383	10,828
Wind	85	3,488	−85	3,488	9,090
Diesel	657	443	−106	995	4,345
Coke	507	0	0	507	3,711
Biomass	325	0	0	325	815
Other fossil fuels	152	0	0	152	307

Source: Authors.
Note: Figures include public service and self-supply.

The analysis of MEDEC interventions that generate, use, or save electricity (that is, all interventions in the electricity sector plus a number of interventions in the stationary energy end-use, oil and gas, and agriculture and forestry sectors) was carried out based on the following assumptions:

- MEDEC interventions (including generation and efficiency) replace baseline production capability. Net generation in the MEDEC scenario is therefore equal to the baseline scenario minus the total energy saved in electricity end-use efficiency interventions.
- Every MEDEC intervention substitutes 86 percent of coal-based generation (supercritical technology) and 14 percent of natural gas generation (combined-cycle technology). The exceptions to this rule are cogeneration in industry and cogeneration in Pemex, which substitute 100 percent natural gas–based generation (because they are considered as efficient ways to use natural gas, not fuel-substitution interventions).
- The total coal capacity displaced in the MEDEC scenario is equal to all new coal power plants foreseen in the baseline scenario (except the 678 MW Carboeléctrica del Pacífico power plant, in the state of Guerrero, which will begin operations in 2010).

Forgone costs in coal and natural gas include investment costs (proportional to new electricity generation or savings), operations and maintenance costs, fossil energy costs, and environmental externalities costs (not included in the reported figures).

The analysis of electricity interventions recognizes the fact that 1 MWh of electricity saved in the distribution grid implies more than 1 MWh saved in generation, because of energy losses in transmission and distribution. Distribution loss factors of 1.012 subtransmission voltage, 1.042 primary voltage, and 1.067 secondary voltage were used (data for Mexico were not available; these data are from Southern California Edison [2008]).

Windpower

- Program definition: Install a power-generation capacity of 10,800 MW
- Project duration: 22 years (21-year lifetime plus 1 year planning and construction)
- Plant factor: 30 percent
- Own consumption: 0
- Investment costs: $1,336,311/MW
- Fixed operations and maintenance costs: $27,458/year/MW
- Variable operations and maintenance costs: 0
- Investment profile: Single year
- Cost factors for externalities (life-cycle analysis): SO_2: $0.003/MWh (gross); sulphates: $0.224/MWh (gross); PM10: $0.094/MWh (gross); NO_X: $0.043/MWh (gross)

Small Hydropower

- Program definition: Install a power-generation capacity of 2,750 MW
- Project duration: 31 years (30 years lifetime plus 1 year planning and construction)
- Plant factor: 45 percent
- Own consumption: 0
- Investment costs: $2,669,523/MW
- Fixed operations and maintenance costs: $36,161/year/MW
- Variable operations and maintenance costs: $4.329/MWh (gross)
- Water use: 12,028 m³/MWh (gross)
- Investment profile: Single year
- Hydropower water cost: $0.00030/m³ (opportunity costs of using the water in other applications)
- Cost factors for externalities (life-cycle analysis): sulphates: $0.023/MWh (gross); PM10: $0.010/MWh (gross)

Geothermal Power

- Program definition: Install a power-generation capacity of 7,500 MW
- Project duration: 33 years (30 years lifetime plus 3 years planning and construction)
- Plant factor: 90 percent
- Own consumption: 0
- Investment costs: $2,803,515/MW
- Fixed operations and maintenance costs: $146,269/year/MW
- Variable operations and maintenance costs: $0.041/MWh (gross)
- Cooling water usage: 0.10 m³/MWh (gross)
- Geothermal steam consumption: 19.29 GJ/MWh (gross)
- Geothermal steam cost, levelized: $1.922/GJ
- Percentage of steam cost that corresponds to exploration and other initial investments: 85 percent
- Cost factor for externalities: 0
- Investment schedule: Year −3: 3 percent; year −2: 60 percent; year −1: 38 percent

Biogas

- Program definition: Install a power-generation capacity of 930 MW
- Project duration: 22 years (21 years lifetime plus 1 year planning and construction)

It is assumed that in the baseline scenario, landfill biogas is captured and burned. Therefore the reduction in methane emissions is not accounted for, and landfill costs are not included (biogas is considered to be available for free).

- Plant factor: 80 percent
- Own consumption: 0
- Investment costs: $3,226,104/MW

- Fixed operations and maintenance costs: $16,613/year/MW
- Variable operations and maintenance costs: $8.039/MWh (gross)
- Investment profile: Single year
- Distribution loss factor: Subtransmission voltage
- Cost factors for externalities: SO_2: $0.007/MWh (gross); sulphates: $0.456/MWh (gross); PM10: 0; NO_x: $0.539/MWh (gross)

Utility Efficiency

- Program definition: Substitute several auxiliary equipments in power plants, transmission, and distribution
- Project duration: 30 years
- Exchange rate: 10.8 pesos/$
- Barrel of oil equivalent per GWh: 2.4 BOE/MWh (PAESE)
- Costs and savings: See table C.5

Table C.5 Costs and Savings for Utility Efficiency Actions

Item	Total PAESE investment (millions of pesos per unit)	PAESE program annual savings (BOE per unit)	Number of units in program	Life-time (years)
Power plants				
Energy audit[a]	1	n.a.	66	30
Variators	2.1	2,637	198	10
Compressors	3	2,160	264	10
Ventilators	2	12,125	132	10
Vapor-vapor generators	2	6,900	132	5
Controllers	2	50,000	132	10
Burners	2	50,000	132	5
Combustion control with viscosity meters	2	19,400	66	15
Transmission and distribution				
Power temperature control for substations	0.08	54.6	660	30
Transformer substitution	0.075	20.8	14,520	30

Source: PAESE.

Note: Total number of units in program and lifetime of assets is based on expert opinion. BOE = barrel of oil equivalent; n.a. = not applicable.

a. The energy audit would identify potential improvements, including through improved routine maintenance and some capital investments, and would specify the energy savings to be gained from these actions.

Oil and Gas Sector

Gas Leakage Reduction

- Program definition: Reduce leakage of natural gas by replacing seals on 46 natural gas compressors
- Project duration: 25 years

- Emissions per compressor: 38.29 million ft³/year without project, 6.22 million ft³/year with project (PGPB 2006)
- Upstream emissions: Not included
- Investment cost of dry seals per compressor: $444,000

Cogeneration in Pemex

- Program definition: Install a cogeneration capacity of 3,690 MW
- Project duration: 33 years (30 years lifetime plus 3 years planning and construction)

Without project

- Self-supply capacity in Pemex: 2,130 MW (SENER 2008c)
- Plant factor: 50 percent (CRE data)
- Fuel to electricity efficiency: 15 percent
- Operations and maintenance costs: Same as for modern cogeneration plant (fixed cost: $29,050/year/MW; variable cost: $0.368/MWh [gross])
- Current fuel to heat (boiler) efficiency: 35 percent
- Boiler operation costs: $0.200/GJ fuel

Cogeneration assumptions

- Fuel: Natural gas. A number of cogeneration schemes would be fueled by gas coming from the gasification of refinery vacuum residuals; because gasification needs to be carried out for other reasons, its costs are not accounted for here.
- Plant factor: 80 percent
- Own consumption: 2.74 percent
- Investment costs: $1,505,000/MW
- Fixed operations and maintenance costs: $29,050/year/MW
- Variable operations and maintenance costs: $0.368/MWh (gross)
- Cooling water usage: 2.06 m³/MWh (gross)
- Fuel to electricity efficiency: 37 percent
- Fuel to heat efficiency: 42 percent
- Investment schedule: Year −3: 7 percent; year −2: 72 percent; year −1: 20 percent
- Cost factors for externalities: SO_2: $0.001/GJ; sulphates: $0.044/GJ; PM10: $0.011/GJ; NO_X: $0.028/GJ

Refinery Efficiency

- Program definition: Renovation of all six refineries in Mexico
- Project duration: 22 years
- Investment: $2,110,000 for each kB/day of crude oil
- Baseline refinery energy consumption for each kB/day of crude oil: natural gas: 0.252 million ft³/day; diesel: 0.008 kB/day; fuel oil: 0.036 kB/day; electricity: 2.860 GWh/year; vapor: 3.631 t/hr
- Reduction in fuel use: 12 percent
- Reduction in electricity use: 0
- Investment profile: Years −3 to −1: 33 percent/year

- Refinery data and schedule: Salina Cruz (2009): 308 kB/day; Tula (2012): 296 kB/day; Minatitlán (2015): 285 kB/day; Madero (2018): 188 kB/day; Cadereyta (2021): 235 kB/day; Salamanca (2024): 176 kB/day

Stationary Energy End-Use Sectors

A number of assumptions for these sectors are based on estimates by Odón de Buen, energy efficiency expert. The assumptions for the interventions that address the commercial and service sectors (nonresidential buildings) are included in table C.6.

Table C.6 Scope for Energy Savings from Nonresidential Air-Conditioning and Lighting Interventions, by Type of Building

Type of building	Total bldg stock (million m²)	No. of bldgs (thousands)	Bldgs w/ AC (%)	Avg. AC energy (MJ/m²/yr)	Bldgs w/ old AC technology (%)	Avg. lighting energy (MJ/m²/yr)	Bldgs w/ old lighting technology (%)
Warehouses	5	1	50	100.00	80	170.33	75
Hotels	12	13	80	289.94	70	281.04	25
Restaurants	2	10	100	289.94	70	281.04	50
Office buildings	4	8	50	148.34	75	143.79	75
Wholesale and retail properties	15.2	2.1	100	177.18	75	171.75	75
Theaters and recreational facilities	2.8	2	100	226.61	75	219.65	75
Hospitals and health facilities	6	21	100	313.25	75	303.63	75
Schools	121	150	50	48.32	80	187.36	100
Other services	110	200	50	50.00	80	100.00	50

Source: Based on data from NRCan (2007), adjusted for Mexico, and authors' assumptions.
Note: Figures assume 10 hours per day of air-conditioning use for all building types. AC = air conditioning.

Nonresidential Air Conditioning

- Program definition: Install efficient air conditioning in all nonresidential buildings
- Project duration: 30 years (equal to air-conditioning lifetime)
- Demand per ton of standard air conditioning: 1.7 kW
- Demand per ton of efficient air conditioning: 0.9 kW
- Cost per ton of efficient air conditioning: $1,140
- Time for implementation of full program: 10 years
- Air conditioning lifetime: 30 years

Nonresidential Lighting

- Program definition: Bring forward by 10 years installation of efficient lighting in all nonresidential buildings
- Project duration: 22 years
- Power of standard equipment (T12 with electromagnetic ballast): 0.192 kW/device
- Power of efficient equipment (T8 with electronic ballast): 0.09 kW/device
- Cost per efficient unit: $55/device
- Time for implementation of full program: 10 years
- Hours per day of lighting use, for all building types: 12
- Applicable distribution loss factor: Primary low voltage

Street Lighting

- Program definition: Bring forward by 10 years the substitution of all street lighting lamps in Mexico by high-pressure sodium lamps
- Project duration: 22 years
- Project schedule: All lamps replaced in 10 years
- Operating hours per year: 4,380
- Energy consumption in 2006: 4,303,000 MWh
- Other technology assumptions: See table C.7

Table C.7 Technology Assumptions for Street Lighting

Application/type of lamp	Power (watts)	Lumens	Estimated use (% of total)	Lamp cost ($)
Main streets				
Mercury vapor	400	23,000	10.00	n.a.
Halogen (iodine-quartz)	1,000	21,000	2.50	n.a.
High-pressure sodium	250	28,000	12.50	84.60
Secondary main streets				
Mercury vapor	250	13,000	7.50	n.a.
Fluorescent	215	14,800	2.50	n.a.
Mixed light	500	14,750	2.50	n.a.
High-pressure sodium	150	16,000	12.50	20.70
Neighborhood streets				
Mercury vapor	125	6,300	12.50	n.a.
Incandescent	300	6,300	6.25	n.a.
Halogen (iodine-quartz)	300	6,000	3.13	n.a.
Fluorescent	85	5,250	3.13	n.a.
High-pressure sodium	70	6,300	25.00	19.10

Source: Authors.
Note: n.a.= not applicable.

Industrial Motors

- Program definition: Accelerate substitution of old, high-usage motors and leapfrog to high-efficiency motors in Mexican industry
- Project duration: 30 years
- Project schedule: All motors are substituted in 7 years
- Demand factor for high-usage motors to be included in program: 5,000 hours/year
- Efficiency before motor substitution: 86 percent

Without project assumptions

- Cost of standard motor: $25/HP (market survey)
- Efficiency of new standard motor: 90 percent (current standard)
- Period over which baseline substitution would take place: 15 years

Project assumptions

- Applicable distribution loss factor: Subtransmission voltage
- Cost of high-efficiency motor: $57.50/HP (market survey)
- Efficiency of high-efficiency motor: 96 percent
- Reported use of electricity in Mexican industry in 2007: 106,633 GWh/year (SENER 2008c)
- Consumption of electricity reported as "industrial" that actually corresponds to service sector: 22,000 GWh/year
- Annual growth of electricity consumption in Mexican industry 3.50 percent (SENER 2008c)
- Percentage of electricity used in motors in industry: 70 percent
- Average demand factor for all industrial motors: 4,000 hours/year
- Percentage of total motor capacity included in program (meets program criteria): 70 percent

Cogeneration in Industry

- Program definition: Install a cogeneration capacity in industries of 6,720 MW
- Project duration: 33 years (30 years lifetime plus 3 years planning and construction)

Without project assumptions

- Boiler efficiency: 75 percent
- Boiler operation costs: $0.200/GJ fuel

Cogeneration assumptions

- Cogeneration plant substitutes natural gas combined-cycle centralized generation
- Fuel: Natural gas
- Plant factor: 80 percent
- Own consumption (by the cogeneration plant itself): 2.74 percent
- Investment costs: $1,505,000/MW
- Fixed operations and maintenance costs: $29,050/year/MW
- Variable operations and maintenance costs: $0.368/MWh (gross)

- Cooling water usage: 2.06m³/MWh (gross)
- Fuel to electricity gross efficiency: 35 percent
- Fuel to heat efficiency: 40 percent
- Distribution loss factor: Subtransmission voltage
- Investment schedule: Year –3: 7 percent; year –2: 72 percent; year –1: 20 percent
- Cost factors for externalities: SO_2: \$0.001/GJ; sulphates: \$0.044/GJ; PM10: \$0.011/GJ; NO_x: \$0.028/GJ

Bagasse Cogeneration

- Program definition: Install efficient cogeneration plants in 55 sugar factories
- Project duration: 27 years

Background assumptions

- Sugarcane consumption per factory: 1 Mt sugarcane/year
- Bagasse yield ratio: 0.3 t bagasse/t sugarcane
- Bagasse heat value (50 percent humidity): 8 GJ/t
- Electricity and mechanical energy consumption by sugar factory: 0.04 MWh/t sugarcane
- Current share of electricity purchased from the grid: 25 percent
- Current fuel oil consumption: 8 l/t sugarcane
- Factory working days: 155 days/year

Project assumptions

- Investment in boilers, power plant, transformers: \$2.5 million/MW
- Investment schedule: 2 years, 50 percent each
- Electricity efficiency of cogeneration unit: 20 percent (assuming a 62-bar, 2-bar back-pressure, system)
- Fuel oil consumption with cogeneration project: 0 l/t sugarcane
- Operations and maintenance costs and externalities related to local emissions are assumed to be the same with and without the project
- Distribution loss factor: Subtransmission voltage

Residential Air Conditioning

- Program definition: Provide thermal insulation and accelerate substitution of residential air conditioners in 1 million high-consumption households
- Project duration: 30 years

Project assumptions

- Cost of new device \$488 (IIE 2006)
- Air-conditioning lifetime: 15 years
- Applicable distribution loss factor: Secondary low voltage

Without project assumptions

- Energy consumption before substitution: 4,000 kWh/year
- Period over which baseline substitution would take place: 15 years

- Consumption after substitution to standard-compliant equipment: 2,800 kWh/year

With project assumptions
- Energy consumption with new device plus thermal insulation: 700 kWh/year
- Cost of thermal insulation: $1,200

Program assumptions
- Total number of households in program: 1 million (based on INEGI data)

Residential Lighting

- Program definition: Replace the most important lamps in 80 percent of households in Mexico by fluorescent lamps
- Project duration: 10 years

Market assumptions
- Current annual incandescent bulb sales: 210 million bulbs (CONUEE)
- Total bulbs per household: 8 (FIDE)
- Number of existing fluorescent lamps: 35 million bulbs (authors' assumption)

A model that divides household lamps into four categories with different hours per day of use was developed, fitting the above assumptions. Technology assumptions appear in table C.8.

Table C.8 Technology Assumptions for Residential Lighting

Lamp type	Incandescent	Fluorescent
Laboratory lifetime (hours)	1,000	8,000
Reduction in lifetime because of voltage variations and other factors (%)	25	25
Lamp cost ($)	0.50	3.00
Efficacy (lumens/watt)	16	60

Source: Authors.

Program assumptions
- Number of electricity paying households: 28.2 million
- Number of nonpaying households: 1 million
- Replacement program will replace lamps used at least: 1 hr/day
- Program coverage: 80 percent of households
- Applicable distribution loss factor: Secondary low voltage

Residential Refrigeration

- Program definition: Accelerate substitution of old residential refrigerators in Mexico
- Project duration: 30 years
- Applicable distribution loss factor: Secondary low voltage

Without project assumptions

- Energy consumption: 0.850 MWh/year (older refrigerators have higher consumption, of about 1.050 MWh/year, but a large number comply with the 1996 standard)

With project assumptions

- Energy consumption: 0.369 MWh/year
- Cost of new 9ft^3 refrigerator: $203 (based on market survey)
- Refrigerator lifetime: 15 years

Program assumptions

- Refrigerators to be substituted by program: 10 million refrigerators (based on INEGI data)
- Number of years to achieve target: 5
- Number of years for substitution in the baseline scenario: 20

Solar Water Heating

- Program definition: Install by 2030 solar water heaters in 60 percent of existing (2009) households and 65 percent of new households
- Project duration: 22 years

Assumptions for new and existing households

- Hot water consumption: 75 l/day/person
- Household occupancy: 4 people/household (CONAPO 2006)
- Required temperature increase: 25°C
- Size of solar water heater: 4 m^2
- Lifetime: 22 years
- Solar radiation: 18 MJ/day/m^2 (PROCALSOL 2007)
- Solar water heater efficiency: 50 percent
- Gas water heater efficiency: 60 percent
- In households with solar water heating, gas is used as a backup to supply 10 percent of water heating energy needs

Assumptions for existing households

- Cost of solar water heater: $1,050
- Installation cost: $262

Program assumptions

- Number of households in 2009: 27.5 million
- Share of 2009 households that will have water heating (of any kind) in 2030: 60 percent; out of this set of households, those that will have solar water heater in 2030 are in
  - baseline: 1 percent
  - intervention: 60 percent

Assumptions for new households

- Cost of solar water heater for new households: $875
- Installation cost: $175

Program assumptions

- Number of households in 2030: 39 million (CONAPO 2008)
- Share of new households with water heating (of any kind) in 2009–30: 80 percent; out of this set of households, those that will have solar water heater in 2030 are in
  - baseline: 10 percent
  - intervention: 65 percent

Improved Cookstoves

- Program definition: Install improved cookstoves in all households with traditional biomass open fires
- Project duration: 24 years

Project assumptions

- One-time investment in training and promotion: $34/stove
- Investment: $84.45/stove
- Stove lifetime: 4 years
- Adoption rate: 60 percent
- Annual monitoring and administration costs: $16/stove
- Annual maintenance cost: $14/stove
- Annual open-fire fuelwood consumption (dry matter): 4.2 TDM/stove
- Savings factor for improved cookstove: 50 percent
- Emission factor open fire: 2 t CO_2e/TDM (Johnson and others 2009)
- Emission factor improved cookstove: 1.62 t CO_2e/TDM (Johnson and others 2009)
- Emission factors include non-Kyoto gases
- Fuelwood cost: $26.25/TDM (García-Frapolli and others forthcoming)
- Effective time savings per day because of use of the improved stove: 0.25 hours/stove/day (García-Frapolli and others forthcoming)
- Benefits from reduced health damages and environment protection (externalities): $341.64/stove/year (García-Frapolli and others forthcoming)

Program assumptions

- Number of fuelwood-using households in Mexico at 2030 in the baseline scenario: 3,878,070
- Fuelwood productivity: 2.9 TDM/hectare/year

Transport Sector

- Baseline assumptions: See table C.9
- Impact on unpaid time costs (time lost by society because of congestion): 0.030 hour/km of total urban distance
- Urban area, 2009: 11,854 km²
- Annual growth of urban area: 0.89 percent

Table C.9 Baseline Assumptions for Transport Sector

Item	Gasoline vehicles	Diesel vehicles
Vehicle fleet, 2009 (millions)	24.4	1.27
Vehicle fleet, annual growth (%)	5	4
Average efficiency, 2009 (km/liter)	7.87	3.08
Average annual increase in efficiency (%)	1.64	0.23
Total average distance, 2009 (km/year/vehicle)	14,167	59,416
Urban distance as percent of total average distance, 2009	92.5	34.84
Externalities ($ per liter of fuel used in urban areas)	0.04	0.06

Source: Authors, based on assumptions by Centro de Transporte Sustentable de México, A.C.

Bus System Optimization

- Program definition: Redesign all feeder mass transit lines in Mexico and make institutional changes (main axis lines are covered by the BRT intervention)
- Project duration: 24 years
- Minibus (small passenger bus) mileage: 73,000 km/year/bus
- Redundancy percentage without project: 34 percent (according to transit plan for city of Querétaro)
- Minibus (small passenger bus) efficiency: 2.9 km/l
- Minibus (small passenger bus) lifetime: 12 years
- Minibus (small passenger bus) cost: $40,000/minibus
- Annual maintenance costs per minibus: $1,034
- Driver salary (two drivers per bus): $556/month/driver
- The intervention assumes no new investment costs, only forgone investments
- Baseline assumption for number of minibuses in 2030: 1.1 million

Urban Densification

- Program definition: Reduce annual urban area growth from 0.89 percent to 0.4 percent
- Project duration: 22 years
- Area growth rate: 45 percent of baseline rate
- Lag in time to obtain results: 3 years
- Infrastructure cost/km^2: $4,088,342 for low-density cities, $4,566,235 for high-density cities (Transit Cooperative Research Program 1998)
- Annual operation costs/km^2: $290,563 for low-density cities, $525,764 for high-density cities (Transit Cooperative Research Program 1998)

This intervention assumes no new investment needed, only forgone investment. Urban area growth reduction involves reduction in urban trip distances proportional to the square root of the urban area, reduction in urban infrastructure and operation costs, and reduction in unpaid time costs, proportional to distances.

Bus Rapid Transit Systems

- Program definition: Establish 122 BRT lines
- Project duration: 24 years
- Length of line: 15 km
- Passengers per line: 125,000 trips/day
- Number of standard buses replaced by one articulated bus: 4
- Number of articulated buses per line: 50
- Average trip length: 11 km
- Cost of infrastructure: $1.8 million/km
- Cost of articulated bus: $300,000/bus
- Cost of standard bus: $120,000/bus
- Articulated bus lifetime: 12 years
- Standard bus lifetime: 12 years
- Standard bus mileage: 73,000 km/year
- Articulated bus mileage: 250 km/day
- Usage factor articulated bus: 300 days/year
- Maintenance costs articulated bus: $0.26/km
- Salary for driver of articulated bus: $741/month
- Annual maintenance costs standard bus: $1,034/bus/year
- Salary for driver of standard bus: $556/month
- Drivers: 2 drivers/bus
- Other assumptions: See table C.10

Table C.10 Assumptions for Vehicles and Passengers before and after BRT Intervention

Vehicle	Trips forgone[a] (%)	Vehicle occupancy (passengers/ vehicle)	Vehicle efficiency (km/l)	Fuel
Private car	10	1.3	9.3	Gasoline
Taxi	6	1.2	10.0	Gasoline
Standard bus	84	27.3	2.3	Diesel
Articulated bus	n.a.	130.0	1.8	Diesel

Source: Authors, based on assumptions by Centro de Transport Sustentable de México, A.C.
Note: n.a. = not applicable.
a. Percentage of BRT passengers traveling by other means before using BRT.

Nonmotorized Transport

- Program definition: Raise proportion of trips by bicycle in Mexican cities to 6 percent by 2030, through the building of cyclepaths
- Project duration: 60 years

Project assumptions
- Cyclepath length: 100 km
- Cyclepath cost: $110,000/km

- Bicycles purchased by users: 200 bicycles/km of cyclepath
- Bicycle cost: $100/bicycle
- Bicycle lifetime: 5 years
- Average trip length: 11 km
- Trips per year in year 2030: 14.8 million trips/year
- Total length of cyclepaths to be built: 37,500 km (based on experience of Portland, Oregon, where similar density of cyclepaths led to 6 percent of trips by bicycle)

Without project assumptions
- Road infrastructure costs: $5 million
- Road infrastructure lifetime: 10 years
- Road maintenance: $400,000/year
- Car cost: $7,500/car
- Car maintenance: $750/year
- Cars forgone: 2 cars/km of cyclepath
- Car lifetime: 12 years
- Other assumptions: See table C.11

Table C.11 Assumptions about Vehicles and Passengers before Nonmotorized Transport Intervention

Vehicle	Trips forgone as result of intervention (%)	Vehicle occupancy (passengers/vehicle)	Vehicle efficiency (km/l)	Fuel
Buses	62.4	15	2.3	Diesel
Cars	29.2	1.3	9.3	Gasoline
Motorcycles	5.2	1	15	Gasoline
Taxis	3.1	1.2	10	Gasoline

Source: Authors, based on assumptions by Centro de Transporte Sustentable de México, A.C., based on experience of Portland, Oregon.

Fuel Economy Standards

- Program definition: Establish fuel economy standards for cars, sport-utility vehicles, and light-duty vehicles in Mexico
- Project duration: 30 years
- Efficiency increase: Standards apply from 2011; exponential growth until 2015, then linear growth (figure C.1)
- Vehicle lifetime: 15 years
- Additional costs calculated from studies carried out by ARB (2009)

Inspection and Maintenance in 21 Large Cities

- Program definition: Implement inspection and maintenance scheme in Mexico's 21 largest cities (excluding the metropolitan area of Mexico City, where an inspection program is already operational), with a one day a week restriction for older vehicles

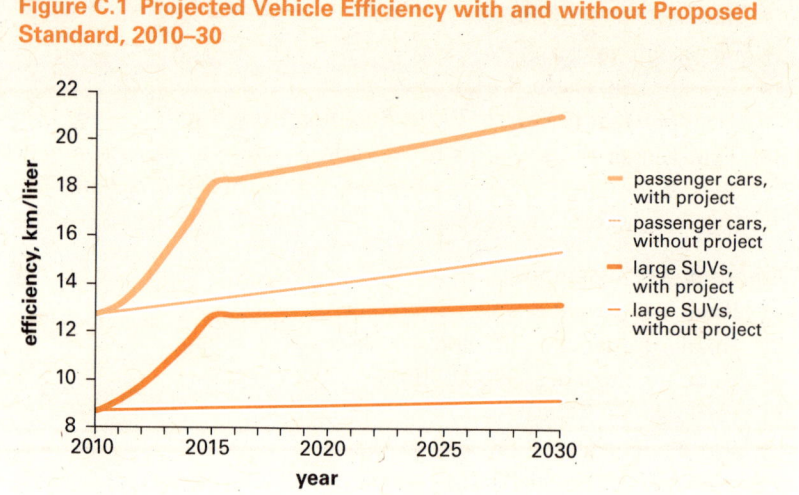

Figure C.1 Projected Vehicle Efficiency with and without Proposed Standard, 2010–30

Source: Authors, based on assumptions by Centro de Transporte Sustentable de México, A.C.

- Project duration: 22 years
- Percentage of total fleet in 21 large cities: 41
- Percentage of gasoline vehicles in large cities that would be subject to inspection and maintenance, 2009: 95.65 (the remaining percentage corresponds to other vehicles such as motorcycles)
- Annual change factor for above percentage: –0.9974
- Percentage of vehicles with one day a week restriction, year 1: 70
- Annual change factor for the percentage: –0.9833
- Percentage of distance reduction for restricted vehicles: 23 percent
- Inspection cost: $46/year/vehicle
- Additional maintenance cost: $55/year/vehicle
- Labor share of inspection and maintenance costs: 40 percent

Border Vehicle Inspection

- Program definition: Inspect second-hand imported vehicles in order to ensure their compliance with national standards on emissions of criteria pollutants
- Project duration: 22 years
- Estimated number of second-hand imported vehicles, 2009: 890,000
- Estimated annual growth for the amount of second-hand imported vehicles: 4 percent
- Estimated percentage of second-hand vehicles that would fail national standards: 20 percent
- Remaining lifetime for imported second-hand vehicles: 8 years
- Costs (assumed to be incurred only by imported vehicles going through inspection):

- Additional maintenance: $55/vehicle
- Inspection: $92/vehicle
- Labor share of costs: 40 percent

Road Freight Logistics

- Program definition: Substitute all single-man owned trucks in the country with freight enterprises or cooperatives. A single company with 80 trucks is assumed to provide the same service as 100 single-man owned trucks as a result the reduction in empty trips.
- Project duration: 24 years
- Truck efficiency: 3.4 km/l
- Truck mileage: 70,000 km/year
- Truck cost: $300,000/truck
- Intervention assumes that there is no new investment; only forgone investment
- Truck life: 12 years
- Truck maintenance: $20,000/truck/year
- Truck driver salary: $741/month
- Drivers: 2/truck
- Management costs for enterprise: $1.5 million/year
- Number of single-man owned trucks to be substituted: 1 million

Railway Freight

- Program definition: Move 37 percent of total long-distance freight by rail by 2030
- Project duration: 22 years
- Road and railway freight assumptions for a 600 km trip: See table C.12
- Current long-distance freight: 321 billion t × km/year
- Current railway share 7.6 percent
- Long-distance freight expected in 2030: 658 billion t × km/year
- Expected railway share without project in 2030: 7.6 percent
- Expected railway share with project in 2030: 37.0 percent

Table C.12 Road and Railway Freight Transport Assumptions

Item	Road	Railway
Capacity (tons)	30	3,000
Load factor (percent)	70	100
Fuel efficiency (km/liter)	1.20	0.033
Operations and maintenance costs excluding fuel ($/trip)	900	90,000

Source: Authors.

Agriculture and Forestry Sector

General assumptions for the sector are as follows:

- Long-distance transport assumptions (for biomass fuels):
  - Truck capacity: 16 tons/truckload
  - Fixed transport cost: $92.59/truckload
  - Variable transport costs excluding diesel: $0.38/km
  - Share of labor cost in fixed and variable transport costs: 50 percent
  - Specific diesel consumption: 0.33 l/km
  - Annual deforestation rate in the baseline: 0.5 percent
  - Annual degradation rate in the baseline: 0.7 percent
  - Emissions from deforestation: 143.9 t CO_2/hectare
  - Emissions from degradation: 28.3 t CO_2/hectare (degradation occurs over several years; it is assumed here to occur in a single year in order to simplify the model)

Biomass Electricity

- Program definition: Install a biomass-fired power-generation capacity of 5,000 MW
- Project duration: 26 years

Power plant assumptions

- Typical power plant size: 24 MW
- Plant factor: 80 percent
- Own consumption: 5 percent
- Gross efficiency: 20 percent
- Investment costs: $2.25 million/MW (Martin 2008)
- Direct labor cost: $5.37/MWh (gross) (based on data from CBC 2008)
- Management and administration: $0.54/MWh (gross) (CBC 2008)
- Maintenance: $4.03/MWh (gross) (CBC 2008)
- Insurance: $3.76/MWh (gross) (CBC 2008)
- Variable operations and maintenance costs:
  - Purchases: $1.07/MWh (gross) (CBC 2008)
  - Ash disposal: $0.54/MWh (gross) (CBC 2008)
  - Other operation expenses: $0.75/MWh (gross) (CBC 2008)
  - Cooling water usage: 2.00 m³/MWh (gross)
  - Investment schedule: Year −3: 10 percent; Year −2: 40 percent; Year −1: 50 percent
  - Cost factors for externalities: SO_2: $0.010/MWh (gross); sulphates: $0.669/MWh (gross); PM10: $0.054/MWh (gross); NO_x: $1.227/MWh (gross)

Biomass production and forest management data

- Deforestation and degradation rates with project: 0
- Fuelwood high heat value: 19 GJ/TDM (De Jong and Olguín-Álvarez 2008)

- Fuelwood (cordwood) productivity: 2.9 TDM/hectare/year (De Jong and Olguín-Álvarez 2008)
- Forest management costs, every 10 years: $35/hectare
- Timber productivity: 1.3 TDM/hectare/year
- Timber stumpage price: $92.59/TDM
- Fuelwood harvesting costs (roadside fuelwood cost): $26.24/TDM
- Percent of labor cost in harvesting: 65
- Fuelwood handling and chipping: $8.50/TDM
- Harvestable area (area around the power plant available to be harvested): 30 percent
- Emissions from fuelwood combustion: 0.050 t CO_2e/MWh

Fuelwood Co-firing

- Program definition: Retrofit the six units of the Petacalco power plant, with a combined capacity of 2,100 MW, so that they are fired by a mix of 80 percent coal and 20 percent biomass
- Project duration: 22 years

Power plant assumptions without project

- Capacity before retrofitting: 350 MW
- Plant factor: 90 percent
- Own consumption: 7.2 percent
- Fixed operations and maintenance costs: $34,619/year/MW
- Variable operations and maintenance costs: $0.198/MWh (gross)
- Cooling water usage: 2.79 m³/MWh (gross)
- Gross efficiency: 40.81 percent
- Cost factors for externalities: SO_2: $0.261/MWh (gross); sulphates: $17.440/MWh (gross); PM10: $1.995/MWh (gross); NO_X: $1.798/MWh (gross)

Power plant assumptions with project

- Plant factor: 87 percent
- Biomass use: 20 percent
- Own consumption: 7.2 percent
- Retrofitting investment costs: $260,000/MW
- Fixed operations and maintenance costs: $34,619/year/MW
- Variable operations and maintenance costs: $0.198/MWh (gross)
- Cooling water usage: 2.79 m³/MWh (gross)
- Gross efficiency: 37.81 percent
- Investment schedule: Single year
- Cost factors for externalities: SO_2: $0.061/MWh (gross); sulphates: $0.000/MWh (gross); PM10: $0.381/MWh (gross); NO_X: $1.045/MWh (gross)
- Greenhouse gas emissions from fuelwood combustion: 0.050 t CO_2e/MWh

Biomass production and forest management data

- Deforestation and degradation rates with project: 0
- Fuelwood high heat value: 19 GJ/TDM (De Jong and Olguín-Álvarez 2008)
- Fuelwood productivity: 2.9 TDM/hectare/year (De Jong and Olguín-Álvarez 2008)
- Forest management costs, every 10 years: $35/hectare
- Timber productivity: 1.3 TDM/hectare/year
- Timber stumpage price: $92.59/TDM
- Fuelwood harvesting costs (roadside fuelwood cost): $26.24/TDM
- Percent of labor cost in harvesting: 65
- Fuelwood handling and chipping: $16.00/TDM
- Harvestable area: 30 percent

Charcoal Production

Program definition

- Part A: Meet 75 percent of industrial coke demand in Mexico with charcoal
- Part B: Improve charcoal production for urban (residential and commercial) consumption by ensuring sustainable forest management and substituting traditional kilns by improved kilns for 70 percent of charcoal production

Part A Project assumptions

- Project duration: 31 years
- Module size: 500 hectares
- Module is divided into 10 equal parts, called coupes. Annual coupe size: 50 hectares (exploitation will be made in 10-year cycles; every year a new coupe is exploited)
- Average standing stock: 150 m³/hectare
- Average allowed cut per coupe: 45 m³/hectare
- Area per kiln: 3.125 hectare
- Dry matter contents: 65 TDM/m³ wood
- Wood to charcoal conversion (improved kilns): 0.3 t charcoal/TDM
- Large-scale charcoal production costs: See table C.13
- Operation costs: $162/t charcoal
- Share of labor cost in operation costs: 80 percent
- Average long-distance transport: 400 km
- Chainsaws: 8 chainsaws/module
- Liters of gasoline per day per chainsaw: 3 l/day/chainsaw
- Chainsaw use: 200 days/year
- Small local truck capacity: 5 m³/trip
- Gasoline per trip of small truck: 3 l/trip
- Non–CO_2 charcoal kiln emissions: 1.108 t CO_2e/t charcoal (Pennise and others 2001)
- Coke replacement coefficient: 1.00 t charcoal/t coke

Table C.13 Large-Scale Charcoal Production Costs

Production costs	Cost ($)	Lifetime (years)
Preparation of forest management program (1 module, 500 hectare)	9,259	10
Firebreaks and roads construction		
3,000 meters per coupe first year	27,778	1
3,000 meters per coupe following years	20,833	1[a]
Kilns	26,667	2[a]
4-tonner second-hand truck	4,630	3
Chainsaws	5,926	1

Source: Estimates by charcoal experts Enrique Riegelhaupt and Tere Arias.
a. Consecutive coupes are adjacent and can use roads built earlier.

Biomass production and forest management data
- Deforestation and degradation rates with project: 0
- Forest management costs, every 10 years: $35/hectare
- Module timber productivity: 1.3 TDM/hectare/year
- Timber stumpage price: $92.59/TDM

Program assumptions
- National coke demand 2009: 3.29Mt/year
- National coke demand forecast for 2031: 9.5 Mt/year

Part B project assumptions
- Production of charcoal per improved kiln: 54 t charcoal/year/kiln (equivalent to production of one improved kiln)
- Fuelwood cost: $26/TDM fuelwood
- Labor cost: $12/day
- Charcoal price: $185/t
- Traditional and improved charcoal kiln assumptions: See table C.14
- Estimated charcoal demand in 2008: 592,102 t charcoal/year
- Estimated annual growth of charcoal demand: 0.8 percent
- Fuelwood productivity: 2.9 TDM/hectare/year

Forest Management
- Program definition: Place 9 million hectares under forest management
- Project duration: 30 years
- Forest management costs: $35.00/hectare/10 years
- Operations and maintenance costs: $36.50/hectare/year
- Revenue from wood sales: $120/hectare/year
- Revenue without project (opportunity costs): $31.50/hectare/year

Table C.14 Assumptions about Traditional and Improved Charcoal Kilns

Kiln assumptions	Traditional kiln	Improved kiln
Charcoal/wood ratio (t charcoal/TDM) (%)	18	30
Labor days per ton of production	6.00	2.22
Investment per kiln ($)	n.a.	1,980
Operations and maintenance first year per kiln (training and supervision costs) ($)	n.a.	146
Lifetime (years)	n.a.	5
Emissions of CO_2 (t CO_2e/t charcoal)	2.403	1.382
Emissions of other gases (CH_4 and N_2O) (t CO_2e/t charcoal)	1.106	1.108
Percentage of nonrenewability	80	0

Source: Pennise and others 2001; estimates by charcoal experts Enrique Riegelhaupt and Tere Arias.
Note: n.a. = not applicable.

Wildlife Management

- Program definition: Place 30 million hectares under wildlife management
- Project duration: 22 years
- Investment costs: $15/hectare
- Operations and maintenance costs: $36.50/hectare/year
- Revenue from agritourism: $4/hectare/year
- Revenue without project (opportunity costs): $31.50/hectare/year

Payment for Environmental Services

- Program definition: Place 5 million hectares under compensation for environmental services
- Project duration: 22 years
- Investment costs: 0
- Operations and maintenance costs: $35.19/hectare/year
- Revenue with project: 0
- Revenue without project (opportunity costs): $31.50/hectare/year

Afforestation

- Program definition: Afforest 1.5 million hectares
- Project duration: 30 years

Project assumptions

- Investment: $1,120/hectare
- Maintenance: $230/hectare/year (this is considered as investment in the economic analysis because it takes place only during the first five years)
- Opportunity costs: $140/hectare/year
- Harvest factor: 30 percent
- Harvest pattern: First 30 percent harvest at year 10; second 30 percent harvest at year 20; final harvest (100 percent) at year 30

- Stumpage value: $20/m³
- Percentage of carbon contents of harvest that is emitted to the atmosphere: 50 percent

Sequestration data
- Growth: 9.92 m³/hectare/year
- Specific weight: 0.6 TDM/m³
- Carbon contents of dry matter: 0.48 t C/TDM

Reforestation and Restoration

- Program definition: Reforest or restore 4.5 million hectares of forest
- Project duration: 30 years

Project assumptions
- Investment: $1,119.57/hectare
- Maintenance: $229.56/hectare/year (this is considered as investment in the economic analysis because it takes place only during the first five years)

Without project assumptions
- Cattle productivity: 40 kg/hectare/year
- Cattle price per kg (alive): $2
- Expenses as a percentage of gross income: 80 percent

Sequestration assumptions
- Forest growth: 4.71 m³/hectare/year
- Specific weight: 0.6 TDM/m³
- Carbon contents of dry matter: 0.48 t C/TDM

Zero-Tillage Maize

- Program definition: Convert 2.5 million hectares from traditional maize agriculture to zero-tillage maize agriculture
- Project duration: 24 years

Project assumptions
- Opportunity costs: $139/hectare/year (FIRA 2006a, 2006b)
- Technical services: $37/hectare/year (FIRA 2006a, 2006b)
- Administration costs: 15 percent of variable costs (FIRA 2006a, 2006b)
- Tractor costs per hour excluding diesel: $19/hour
- Labor cost: $11/day
- Diesel consumption by tractor: 8.21l/hour
- Maize price: $259/t (market survey)
- Stubble price: $74/t (market survey June 2008: $Mex20 per 25 kg pack)
- Stubble production: 5 t/hectare/year (Etchevers, Tinoco, and Riegelhaupt 2008)
- Baseline and zero-tillage costs: See table C.15
- Cost of additional machinery: $40,000 (Etchevers, Tinoco, and Riegelhaupt 2008)

Table C.15 Baseline and Zero-Tillage Costs

Item	Without project	With project	
		First year	Ensuing years
Tractor time (hours/year/hectare)	17	12.25	8
Labor time (days/year/hectare)	14.5	10.6	8.25
Seed cost ($/hectare/year)	111	111	111
Agrochemical products (herbicides) ($/hectare/year)	106	106	160
Fertilizer ($/hectare/year)	324	324	389
Productivity (t/hectare)	3.20	3.20	Gradually increasing
Annual productivity gain (t/hectare/year)	n.a.	n.a.	0.1
Stubble availability for sale (%)	70	0	0

Source: FIRA 2006a and 2006b; Etchevers, Tinoco, and Riegelhaupt 2008.
Note: n.a. = not applicable.

- Machinery lifetime: 8 years
- Area covered by one machine: 810 hectares/year (60 days/year × 13.5 hectares/day)
- Incorporation of stable organic matter to soil: 0.20 t/hectare/year
- Content of carbon in organic matter: 85 percent

Sugarcane Ethanol

- Program definition: Develop 1.5 million hectares of sugarcane ethanol production
- Project duration: 29 years

Project assumptions

- Typical plant capacity: 85,000 m³ ethanol/year; 550 m³/day
- Lifetime: 25 years
- Unit investment cost: $388/m³ ethanol/year
- Investment profile: Year −4: 9.1 percent; Year −3: 22.7 percent; Year −2: 27.3 percent; Year −1: 40.9 percent
- Factory operations and maintenance: $4.40/t sugarcane
- Labor share: 30 percent
- Ethanol conversion factor: 0.08 m³ ethanol/t sugarcane

Sugarcane costs

- Transport cost: $3.06/t sugarcane (FIRA 2007)
- Field costs: $30/t sugarcane (FIRA 2007)
- Labor share: 40 percent

Energy production

- Electricity generation: 0.08 MWh/t sugarcane
- Electricity consumption: 0.03 MWh/t sugarcane
- Distribution loss factor: Subtransmission voltage

Plantation assumptions

- Yield: 61 t/year/hectare

Sorghum Ethanol

- Program definition: Develop 3 million hectares of sorghum ethanol production
- Project duration: 30 years

Project assumptions

- Typical plant capacity: 165,000 m³ ethanol/year; 0.5 million l/day
- Lifetime: 25 years
- Unit investment cost, including planning: $557/m³ ethanol/year
- Investment profile: Year −5: 9.1 percent; Year −4: 22.7 percent; Year −3: 27.3 percent; Year −2: 18.2 percent; Year −1: 22.7 percent
- Operations and maintenance: $11.42/t sorghum
- Labor share of operations and maintenance costs: 12 percent
- Ethanol conversion factor: 0.36 m³ ethanol/t sorghum
- Transport cost, average 100 km: $8.49/t sorghum
- Field costs: $135/t sorghum
- Labor share of field costs: 18 percent
- By-product sales: DDG yield: 0.333 tons DDG/t sorghum; value: $140/t DDG
- Electricity consumption: 0.0756 MWh/t sorghum
- Distribution loss factor: Primary low voltage
- Natural gas consumption: 0.00835 GJ/t sorghum
- Sorghum high yield: 3.5 t/year/hectare
- Sorghum medium yield: 2 t/year/hectare
- High- and medium-yield surfaces grow up to a total of 3 million hectares

Palm Oil Biodiesel

- Program definition: Develop Mexico's production of palm oil bio-diesel reaching a surface area of 215,000 hectares in 2030
- Project duration: 25 years

Project assumptions

- Plant investment, including planning stage: $12,482,800/plant
- Plant capacity: 37,854 m³/year
- Investment profile: Year −2: 9 percent; year −1: 91 percent
- Fixed operations and maintenance costs: $377,900/year/plant
- Labor share of operations and maintenance costs: 47 percent
- Cost of fresh fruit bunches: $111/t
- Labor share of fresh fruit bunch costs: 45 percent
- Oil yield from fresh fruit bunches: 20.40 percent (www.fedepalma.org)
- Use of other raw materials: See table C.16
- Miscellaneous materials: $153,000/year/plant
- Greenhouse gas emissions from the use of these materials are calculated from IPCC emission factors

Table C.16 Use of Raw Materials in the Production of Biodiesel

Raw material	Needs (t/year/plant)	Unit cost ($/t)
CH_3OH	3,921	278.53
$NaOCH_3$	329	953.54
HCl	273	128.06
NaOH	167	598.80
Water	1,124	1.78

Source: Estimates by Oliver Probst, Instituto Tecnológico y de Estudios Superiores de Monterrey.

- Electricity consumption: 1,008 MWh/year/plant
- Distribution loss factor: Primary low voltage
- Diesel consumption (for transportation): 10,000 GJ/year/plant

By-products
- Kernel: 7,716 t/plant/year, at $150/t
- Glycerine: 3,429 t/plant/year, with no value ($0.00/t)

Plantation data
- Average yield: 16.3 t fresh fruit bunch/year/hectare (INEGI data)
- Production curve reaches maximum of 24 t/hectare/year for high-yield areas and then declines
- Medium-yield areas: 60 percent of high-yield areas

Bibliography

Antonius, A., S. Awerburch, M. Berger, D. Hertzmark, J. M. Huacuz-V., and G. Merino. 2006. *Mexico: Technical Assistance for Long-term Program of Renewable Energy Development.* ESMAP Technical Paper 093, World Bank, Energy Sector Management Assistance Program, Washington, DC. http://tinyurl.com/2jxcuj.

ARB (Air Resources Board). 2009. *Documents from the Air Resources Board of the California Environmental Protection Agency.* http://tinyurl.com/ARBggemv.

Armendáriz, C., R. D. Edwards, M. Johnson, M. Zuk, L. Rojas, R. D. Jiménez, H. Riojas-Rodriguez, and O. R. Masera. 2008. "Reduction in Personal Exposures to Particulate Matter and Carbon Monoxide as a Result of the Installation of a Patsari Improved Cook Stove in Michoacan Mexico." *Indoor Air* 18 (2): 93–105.

Bacon, R., J. Halpern, and R. Boyd. 2004. *Energy Policies and the Mexican Economy.* World Bank, Energy Sector Management Assistance Program, Washington, DC. http://tinyurl.com/2ohouh.

Boyd, R., and M. E. Ibarrarán. 2008. "Extreme Climate Events and Adaptation: An Exploratory Analysis of Drought in Mexico." *Environmental and Development Economics.* Cambridge University Press, Cambridge. http://tinyurl.com/doi4956.

CBC (California Biomass Collaborative). 2008. *Cost of Energy Calculator.* http://tinyurl.com/UCDcec.

Center for Clean Air Policy-Europe, Centre for European Policy Studies, Climate Change Capital, Centre for European Economic Research, and Institute for Sustainable Development and International Relations. 2008.

Sectoral Approaches: A Pathway to Nationally Appropriate Mitigation Actions. Interim report. http://tinyurl.com/SectApp.

CFE (Comisión Federal de Electricidad). 2008a. *Costos y parámetros de referencia para la formulación de proyectos de inversión en el sector eléctrico: Generación.* Mexico City.

————. 2008b. *Programa de Obras e Inversiones del Sector Eléctrico 2008-2017.* Mexico City. http://tinyurl.com/poise2017.

Christensen, J. H., B. Hewitson, A. Busuioc, A. Chen, X. Gao, I. Held, R. Jones, R. K. Kolli, W. T. Kwon, R. Laprise, V. Magaña-Rueda, L. Mearns, C. G. Menéndez, J. Räisänen, A. Rinke, A. Sarr, and P. Whetton. 2007. "Regional Climate Projections in Climate Change 2007: The Physical Science Basis." Contribution of Working Group I to the *Fourth Assessment Report of the Intergovernmental Panel on Climate Change,* 847–94. Cambridge: Cambridge University Press. http://tinyurl.com/ar4wg1ch11.

CICC (Comisión Intersecretarial de Cambio Climático). 2009. *Programa Especial de Cambio Climático.* Preliminary version for public consultation. http://tinyurl.com/peccvcp.

CIEco (Centro de Investigaciones en Ecosistemas). 2008. *Análisis integrado de las tecnologías, el ciclo de vida y la sustentabilidad de las opciones y escenarios para el aprovechamiento de la bioenergía en México. Reporte final.* http://tinyurl.com/lcabmex.

CONAE (Comisión Nacional para el Ahorro de Energía). 2002. *Programas estatales de minihidráulica.* http://tinyurl.com/prminih.

CONAFOR (Comisión Nacional Forestal). 2001. *Programa estratégico forestal para México 2025.* http://tinyurl.com/PEF2025.

CONAPO (Consejo Nacional de Población). 2006. *Proyecciones de la población de México 2005–2050.* http://tinyurl.com/PobMex50.

CONUEE (Comisión Nacional para el Uso Eficiente de la Energía). 2009. *Estrategia integral para el fomento de la cogeneración en México.* Draft.

CTS (Centro de Transporte Sustentable de México). 2009. *México, Estudio para la Disminucion de Emisiones de Carbono (MEDEC) en el Sector Transporte.* Mexico City.

Dargay, J., D. Gately, and M. Sommer. 2007. *Vehicle Ownership and Income Growth, Worldwide: 1960–2030.* http://tinyurl.com/voigww.

de Dinechin, F., and G. Larson. 2007. "Returning Young Mexican Farmers to the Land." *Agricultural and Rural Development Notes* 23 (June). World Bank, Washington, DC. http://tinyurl.com/ARD2307.

de Jong, B. H. J., and M. Olguín-Alvarez. 2008. *Mitigation Potential in the Forestry Sector.* World Bank, Washington, DC.

De la Torre, A., P. Fajnzylber, and J. Nash. 2009. *Low Carbon, High Growth: Latin American Responses to Climate Change.* World Bank, Washington, DC. http://tinyurl.com/WB47604.

Delphi Group, and IIE (Instituto de Investigaciones Eléctricas). 2006. *Reviewing Gaps in Resource Mapping for Renewable Energy in North America.* Commission for Environmental Cooperation. http://tinyurl.com/367s42.

Etchevers, J., J. A. Tinoco, and E. Riegelhaupt. 2008. *Production of Maize under Conservation Tillage.* World Bank, Washington, DC.

FIDE (Fideicomiso para el Ahorro de Energía Eléctrica). 2008. *Resultados.* http://tinyurl.com/FideRes.

FIRA (Fideicomisos Instituidos en Relación con la Agricultura). 2006a. *Cultivos de maíz y sorgo. Analisis de rentabilidad 2005 y costos de cultivo 2006.* Dirección de Consultoría de Agronegocios, Dirección Regional Occidente, Residencia Estatal Guanajuato, August. http://tinyurl.com/FIRAgto.

———. 2006b. *Cultivo de sorgo y maíz amarillo. Análisis de rentabilidad O-I 2005/2006 y Costos de cultivo O-I 2006/2007.* Dirección de Consultoría de Agronegocios, Dirección Regional Norte, Residencia Estatal Tamaulipas, July. http://tinyurl.com/FIRAtamps.

———. 2007. *Caña de azúcar Ingenio Tres Valles. Análisis de Rentabilidad Zafra 2005-2006 y Proyección de la Rentabilidad Zafra 2007–2008.* Dirección de Consultoría de Agronegocios, Dirección Regional del Sur, Residencia Estatal Veracruz. http://tinyurl.com/FIRAver.

Galindo, L.M., ed. 2009. *La economía del cambio climático en México. Síntesis.* SEMARNAT and SHCP. http://tinyurl.com/eccm2009.

García-Frapolli, E., C. Armendáriz, V. M. Berrueta, R. D. Edwards, A. Guevara, H. Riojas-Rodríguez, and O. R. Masera. Forthcoming. *Beyond Fuelwood Savings: Valuing the Economic Benefits of Introducing Improved Biomass Cookstoves in the Purhépecha Region, Mexico.*

Greenpeace, and EREC (European Renewable Energy Council). 2008. *Revolución energética: Una perspectiva de energía sustentable para México.* http://tinyurl.com/GPrevene.

Houdashelt, M. N. Helme, and D. Klein. 2009. *Setting Mitigation Goals for Sectoral Programs: A Preliminary Case Study of Mexico's Cement and Oil Refining Sectors.* http://tinyurl.com/smgfsp.

Hondo, H. 2005. "Life Cycle GHG Emission Analysis of Power Generation Systems: Japanese Case." *Energy* 30: 2042–56. http://tinyurl.com/LCEjapan.

IAEA (International Atomic Energy Agency). 2005. *Comparative Assessment of Energy Options and Strategies in Mexico until 2025. Final Report of a Coordinated Research Project 2000–2004.* http://tinyurl.com/2jgbr7.

IEA (International Energy Agency). 2007. *Tracking Industrial Energy Efficiency and CO$_2$ Emissions.* http://tinyurl.com/tieeco2e.

———. 2008a. *Energy Balances of OECD Countries.*

———. 2008b. *Key World Energy Statistics 2008.* http://tinyurl.com/kwes08.

IIE (Instituto de Investigaciones Eléctricas). 2006. *Realización de mediciones de energía eléctrica en viviendas de interés social para el análisis de ahorros energéticos.* http://tinyurl.com/rmeevis.

IMP (Instituto Mexicano del Petróleo). 2005. *Escenarios de emisiones y medidas de mitigación de gases de efecto invernadero en sectores clave (Transporte y Desechos).* http://tinyurl.com/IMPdesec (waste), http://tinyurl.com/IMPtrans (transport).

———. 2006. *Proyección de emisiones por sector y gas (CO$_2$, CH$_4$, N$_2$O, HFC, PFC, SF$_6$) para los años 2008, 2012 y 2030.* http://tinyurl.com/F21412.

INE (Instituto Nacional de Ecología). 2006. *Estudio de evaluación socioeconómica del proyecto integral de calidad de combustibles, México, D.F.*

INIFAP (Instituto Nacional de Investigaciones Forestales, Agrícolas y Pecuarias). 2006. *Obtención de factores de emision nacionales en el sector agricola para disminuir incertidumbre en el inventario nacional de emisiones de gases de efecto invernadero.* http://tinyurl.com/ofensagr.

IPCC (Intergovernmental Panel on Climate Change). 2007. *IPCC Fourth Assessment Report (AR4). Climate Change 2007: The Physical Science Basis.* http://tinyurl.com/AR4psb.

Johnson, M., R. Edwards, C. Alatorre, and O. Masera. 2008. "In-Field Greenhouse Gas Emissions from Cookstoves in Rural Mexican Households." *Atmospheric Environment* 42 (6): 1206–22. http://tinyurl.com/inggefc.

Johnson, M., R. Edwards, A.Ghilardi, V. Berrueta, D. Gillen, C. Alatorre, and O. Masera. Forthcoming. "Quantification of Carbon Savings from Improved Biomass Cookstove Projects." *Environmental Science & Technology.* http://tinyurl.com/qcsibcp.

Komives, K, T. M. Johnson, J. Halpern, J. L. Aburto, and J. R. Scott. 2009. *Residential Electricity Subsidies in Mexico: Exploring Options for Reform and for Enhancing the Impact on the Poor.* http://tinyurl.com/SKU17884.

LAERFTE (Ley para el Aprovechamiento de las Energías Renovables y el Financiamiento de la Transición Energética). 2008. http://tinyurl.com/laerfte.

Martin, J. R. 2008. "Biomass Energy Economics." Paper presented at the Western Forest Economists 43rd Annual Meeting. http://tinyurl.com/JRMbee.

Masera, O. R., A. D. Cerón, and J. A. Ordóñez. 2001. "Forestry Mitigation Options for México: Finding Synergies between National Sustainable Development Priorities and Global Concerns." *Mitigation and Adaptation Strategies for Climate Change: Special Issue on Land Use Change and Forestry Carbon Mitigation Potential and Cost Effectiveness of Mitigations Options in Developing Countries* 6 (3–4): 291–312. http://tinyurl.com/ForMitMex.

Masera, O., and others. 2006. *La bioenergía en México. Un catalizador del desarrollo sustentable*. Red Mexicana de Bioenergía and Comisión Nacional Forestal. http://tinyurl.com/biocds.

Masera, O., N. Rodríguez-Martínez, I. Lazcano-Martínez, L. A. Horta-Nogueira, I. C. Macedo, S. C. Trindade, D. Thrän, O. Probst, M. Weber, and F. Müller-Langer. 2006. *Potenciales y viabilidad del uso de bioetanol y biodiesel para el transporte en México*. Secretaría de Energía, Inter-American Development Bank, and GTZ. http://tinyurl.com/pvubbtm.

McKinsey, and CMM (Centro Mario Molina para Estudios Estratégicos sobre Energía y Medio Ambiente, A.C.). 2009. *Low-Carbon Growth. A Potential Path for Mexico*.

McNeil, M. A. and V. E. Letschert. 2008. *Future Air Conditioning Energy Consumption in Developing Countries and What Can Be Done about It: The Potential of Efficiency in the Residential Sector*. Paper LBNL-63203, Lawrence Berkeley National Laboratory, University of California, Berkeley. http://tinyurl.com/LBNL63203.

Mulás, P., and others. 2005. *Prospectiva sobre la utilización de las energías renovables en México. Una visión al año 2030*. Universidad Autónoma Metropolitana. http://tinyurl.com/psuerm.

Navarro, A. 2000. *Manual práctico de labranza de conservación*. Secretaría de Agricultura, Ganadería y Desarrollo Rural, México.

NRCan (Natural Resources Canada). 2007. *Commercial and Institutional Consumption of Energy Survey 2005*. http://tinyurl.com/cices06.

PECC (Programa Especial de Cambio Climático). 2009. http://tinyurl.com/pecc2009.

Pemex (Petróleos Mexicanos). 2004. *Cogeneración en Pemex*. Powerpoint presentation, May 14.

———. 2008. *Reportes de resultados financieros*. Mexico City. http://tinyurl.com/Pemexres.

Pennise, D. M., K. R. Smith, J. P. Kithinji, M. E. Rezende, T. J. Raad, J. Zhang, and C. Fan. 2001. "Emissions of Greenhouse Gases and Other Airborne Pollutants from Charcoal Making in Kenya and Brazil." *Journal of Geophysical Research* 106 (2): 143–44, 155. http://tinyurl.com/GHGchKB.

Pitty, A. 1997. *Introducción a la biología, ecología y manejo de malezas.* Tegucigalpa, Honduras: Zamorano.

PND (*Plan Nacional de Desarrollo 2007–2012*). 2007. pnd.presidencia. gob.mx.

PROCALSOL (Programa para la Promoción de Calentadores Solares de Agua en México). 2007. http://tinyurl.com/Procalsol.

Rojas, L. A., A. Mora, and H. Rodríguez. 2002. "Efecto de la labranza mínima y la convencional en arroz en la región huetar norte de Costa Rica." *Agronomía Mesoamericana* 13 (2): 111–16. http://tinyurl.com/ Rojas2002.

SAGARPA (Secretaría de Agricultura, Ganadería, Desarrollo Rural, Pesca y Alimentación). 2007a. *Labranza de conservación.* http://tinyurl.com/ LabCons.

———. 2007b. *Programa sectorial de desarrollo agropecuario y pesquero 2007–2012.* http://tinyurl.com/psdaf12.

Sanchez, I. H. Pulido, M. A. McNeil, I. Turiel, and M. della Cava. 2007. *Assessment of the Impacts of Standards and Labeling Programs in Mexico (Four Products).* Paper LBNL-62813. http://tinyurl.com/LB62813.

Schmidt, J., N. Helme, J. Lee, and M. Houdashelt. 2007. "Sector-Based Approach to the Post-2012 Climate Change Policy Architecture." *Climate Policy* 8(2008): 494–515. http://tinyurl.com/103763.

Sci-Tech Encyclopedia. 1997. "Cogeneration." *McGraw-Hill Encyclopedia of Science and Technology.* http://tinyurl.com/l25b86. Cited at www. answers.com/topic/cogeneration.

Secretaría del Medio Ambiente del Gobierno del Distrito Federal. 2008. *Programa de acción climática de la ciudad de México 2008–2012.* http:// tinyurl.com/PACCcm.

SEMARNAP (Secretaría de Medio Ambiente, Recursos Naturales y Pesca), and INE (Instituto Nacional de Ecología). 1997. *México primera comunicación nacional ante la convención marco de las naciones unidas sobre el cambio climático.* http://tinyurl.com/fncmex.

SEMARNAT (Secretaría de Medio Ambiente y Recursos Naturales). 2007. *Estrategia nacional de cambio climático.* http://tinyurl.com/enacc2007. An executive summary in English can be found at http://tinyurl.com/ enaccSummary.

SEMARNAT (Secretaría de Medio Ambiente y Recursos Naturales), and INE (Instituto Nacional de Ecología). 2001. *México segunda comunicación nacional ante la Convención Marco de las Naciones Unidas sobre el Cambio Climático.* http://tinyurl.com/sncmex.

———. 2006a. *Inventario nacional de emisiones de gases de efecto invernadero 1990–2002.* http://tinyurl.com/INEGEI2002.

————. 2006b. *México tercera comunicación nacional ante la Convención Marco de las Naciones Unidas sobre el Cambio Climático.* http://tinyurl.com/tncmex.

SENER (Secretaría de Energía). 2007. *Electricity Sector Outlook 2007–2016.* http://tinyurl.com/meso16 (original version in Spanish: *Prospectiva del Sector Eléctrico 2007–2016*, http://tinyurl.com/PSE2016).

————. 2008a. *Diagnóstico: Situación de Pemex 2008.* www.pemex.com/files/content/situacionpemex.pdf.

————. 2008b. *Prospectiva del mercado de gas natural 2008–2017.* http://tinyurl.com/pgn2017.

————. 2008c. *Prospectiva del Sector Eléctrico 2008–2017.* http://tinyurl/PSE2017.

————. 2008d. *Sistema de Información Energética.* http://tinyurl.com/senerSIE.

Sheinbaum, C., and O. Masera. 2000. "Mitigating Carbon Emissions While Advancing National Development Priorities: The Case of Mexico." *Climatic Change* 47: 259–82. http://tinyurl.com/mcewandp.

SIACON (Sistema de Información Agroalimentaria de Consulta). 2007. http://tinyurl.com/SIACON.

Southern California Edison. 2008. *Distribution Loss Factors, 2008 Averages.* http://tinyurl.com/SCEDLF.

Taylor, R. P., C. Govindarajalu, J. Levin, A. S. Meyer, and W. A. Ward. 2008. *Financing Energy Efficiency: Lessons from Brazil, China, India and Beyond.* World Bank, Washington, DC. http://tinyurl.com/wb42529.

Transit Cooperative Research Program. 1998. *The Costs of Sprawl—Revisited.* TCRP Report 39, Transportation Research Board, National Research Council. http://tinyurl.com/tcrp39.

Troncoso, K., A. Castillo, O. Masera, and L. Merino. 2007. "Social Perceptions about a Technological Innovation for Fuelwood Cooking: Case Study in Rural Mexico." *Energy Policy* 35 (5): 2799–2810. http://tinyurl.com/spatifc.

Vergara, W. 2008. "Climate Hotspots: Climate-Induced Ecosystem Damage in Latin America." LCR Sustainable Development Working Paper 32, World Bank, Washington, DC. http://tinyurl.com/LCRSDW32.

WCD (World Commission on Dams). 2000. *Dams and Development. A New Framework for Decision-making. The Report of the World Commission on Dams.* http://tinyurl.com/wcdreport.

Wingate, M., J. Hamrin, L. Kvale, and C. Alatorre. 2007. *Fostering Renewable Electricity Markets in North America.* Commission for Environmental Cooperation. http://tinyurl.com/2ecyh4.

World Bank. 2006a. *Project Appraisal Document of the Large-Scale Renewable Energy Development Project.* Washington, DC. http://tinyurl.com/2zzoum.

———. 2006b. *Technical and Economic Assessment of Off Grid, Mini-Grid and Grid Electrification Technologies. Annexes.* Energy Unit, Energy and Water Department, Washington, DC. http://tinyurl.com/wb2006.

———. 2008. *Study of Equipment Prices in the Power Sector.* Draft. Washington, DC. http://tinyurl.com/draftdoc1.

Xiaoyu, Yan. 2008. *Life Cycle Fossil Energy Demand ad Greenhouse Gas Emissions in China's Road Transport Sector.* Queen Mary College, University of London. http://tinyurl.com/dffwjv.

Index